AF389394

Soil Carbon and Microbial Processes
in Agriculture Ecosystem

Soil Carbon and Microbial Processes in Agriculture Ecosystem

Editors

Yinglong Chen
Masanori Saito
Etelvino Henrique Novotny

Basel • Beijing • Wuhan • Barcelona • Belgrade • Novi Sad • Cluj • Manchester

Editors
Yinglong Chen
The University of Western
Australia
Perth, Australia

Masanori Saito
Tohoku University
Osaki, Japan

Etelvino Henrique Novotny
Embrapa Soils
Rio de Janeiro, Brazil

Editorial Office
MDPI
St. Alban-Anlage 66
4052 Basel, Switzerland

This is a reprint of articles from the Special Issue published online in the open access journal *Agriculture* (ISSN 2077-0472) (available at: https://www.mdpi.com/journal/agriculture/special_issues/soil_carbon_microbial).

For citation purposes, cite each article independently as indicated on the article page online and as indicated below:

Lastname, A.A.; Lastname, B.B. Article Title. *Journal Name* **Year**, *Volume Number*, Page Range.

ISBN 978-3-0365-9724-9 (Hbk)
ISBN 978-3-0365-9725-6 (PDF)
doi.org/10.3390/books978-3-0365-9725-6

Cover image courtesy of Masanori Saito

Contents

About the Editors

Yinglong Chen

Yinglong Chen is an ARC Future Fellow and Research Professor at the School of Agriculture and Environment, The University of Western Australia. He has accumulated 30 years of research experience, focusing on the structure and functions of plant roots and rhizosphere interactions. Yinglong has developed a novel phenotyping platform for characterizing world-first root trait variability among diverse germplasms in major food crops and their responses to environmental stress. He has published four textbooks, one translated textbook, three edited books, nine book chapters as an invited author, and over 260 journal articles. Yinglong has collaborated with researchers from more than 200 external research units. His research has received wide media coverage, contributing towards 9 out of the 17 global Sustainable Development Goals (SDGs).

Masanori Saito

Masanori Saito is a Professor Emeritus of Tohoku University. He has more than 40 years of research experience in the broad area of soil science and soil microbiology, and is one of the leading scientists focusing on arbuscular mycorrhiza in agro-ecosystems. He contributed to the establishment of the germplasm of arbuscular mycorrhizal fungi for the MAFF, Japan. His research interests extend to the lifecycle assessment (LCA) of agricultural systems, and his work contributed to the implementation of the agro-environmental policy in the MAFF based on the knowledge of LCA. He has published approximately 200 research articles and contributed to the chapters of several textbooks and dictionaries.

Etelvino Henrique Novotny

Etelvino Henrique Novotny holds a bachelor's degree in agronomy from the Federal University of Paraná (1993), a master's degree in agronomy from the Federal University of Paraná (1997), and a doctorate in chemistry from the University of São Paulo (USP) (2002). He conducted his postdoctoral research at the Institute of Physics, São Carlos, University of São Paulo (USP); the University of Limerick, Ireland (Chemical and Environmental Science); and Victoria University of Wellington (School of Chemical and Physical Sciences). He is a senior researcher at the Brazilian Agricultural Research Corporation (Embrapa Soils). He has experience in soil chemistry, chemometrics, and physical chemistry, focusing on spectroscopy.

Editorial

Soil Carbon and Microbial Processes in Agriculture Ecosystem

Masanori Saito [1,*], Etelvino Henrique Novotny [2] and Yinglong Chen [3]

1 Graduate School of Agriculture, Tohoku University, Osaki 989-6711, Japan
2 Embrapa Soils, Rua Jardim Botânico, 1024 CEP, Rio de Janeiro 22460-000, Brazil; etelvino.novotny@gmail.com
3 The UWA Institute of Agriculture, The University of Western Australia, Perth, WA 6009, Australia;
   yinglong.chen@uwa.edu.au
* Correspondence: masanori.saito.b6@tohoku.ac.jp

Citation: Saito, M.; Novotny, E.H.; Chen, Y. Soil Carbon and Microbial Processes in Agriculture Ecosystem. *Agriculture* **2023**, *13*, 1785. https://doi.org/10.3390/agriculture13091785

Received: 7 September 2023
Accepted: 8 September 2023
Published: 9 September 2023

As global warming progresses, concerns also arise regarding the decline in agricultural productivity and soil degradation. On the other hand, the demand for food is increasing as the population grows, and the maintenance and enhancement of soil productivity to support this demand are becoming important global issues.

Soil carbon (C), in particular, plays a crucial role not only in maintaining soil fertility but also as a global C sink. Soil C is a complex product resulting from various biological, physical, and chemical processes, and it is a fragile entity that is vulnerable to inappropriate human activity and global climate change. Soil microbes with efficient C use help reduce C losses and increase C storage. In this context, it is essential to understand the dynamic nature of soil C and microbial processes in agricultural ecosystems.

To facilitate a deeper understanding of the dynamics of soil C and the microbial processes that affect it, this Special Issue focuses on various aspects of C cycling and its related microbial processes in agricultural ecosystems—from the molecular level to regional or global scales. Topics of interest in this Special Issue include the spatiotemporal dynamics of soil C, C balance, characteristics of soil organic C, C dynamics in plant–soil systems, and various approaches to managing soil C maintenance and C sequestration.

Andosol, derived from volcanic ash, is one of the soils with the highest C content in the world. Elucidating the factors that contribute to this C accumulation in soils will provide the basis for the development of soil C sequestration technologies. Howlett et al. [1] investigated the characteristics of soil C accumulation in Japanese Miscanthus grasslands in terms of microtopography and estimated the accumulation rate of C based on δ^{13}C and ^{14}C dating. They found that the C accumulation in C4 grassland was lower than previously thought. On the other hand, Komal et al. [2] investigated the soil C accumulation potential of agroforestry in the eastern region of Pakistan, which is exposed to the major impacts of climate change due to global warming. They found that C storage potential differed among tree species and observed that woody vegetation C stock was comparable to soil C stock. These results highlight the significance of tree-planting agendas on cultivated lands in terms of climate-smart agriculture in Pakistan.

Agricultural activities, especially fertilization, have a significant impact on the balance of elements in the soil, which in turn affects soil C cycling and accumulation. Liu et al. [3] used a geographic information system to determine spatiotemporal changes in C, N, and P in agricultural land around Taihu Lake in China. Due to years of fertilization, N and P accumulated in areas with high agricultural activity. The results suggested that the imbalanced C:N:P stoichiometry, characteristics of agricultural soil, may lead to decoupled C, N, and P biogeochemical cycles, thereby having an adverse impact on soil C sequestration. Matsuoka-Uno et al. [4] showed that the mineralization of organic matter in humus-rich Andosol is greatly enhanced by liming and phosphorus fertilization, shedding light on the importance of fertilizer management for C accumulation in agricultural land, similar to the study by Liu et al. [3].

Various soil amendments have been developed and used to promote soil C accumulation and proper C cycling. Biofertilizers containing *Trichoderma*, known as plant growth-promoting fungi (PGPF), are widely used to maintain crop growth and health. Zhu et al. [5] studied the effect of *Trichoderma*-containing biofertilizer on the mineralization characteristics of soil aggregates. They found that this material not only improves the soil microenvironment but also maintains soil organic matter by retarding mineralization in fine fractions of soil aggregates. Novotny et al. [6] investigated the effect of biochar addition in vermicompost preparation and found that the simultaneous addition of biochar and superphosphate enhanced the humification of compost and resulted in compost of good quality. In the C cycle in agroecosystems, we cannot forget about methane, which is produced by the anaerobic decomposition of C. Wang et al. [7] examined changes in methane emissions from agricultural land in China from 2000 to 2019 based on land-use maps and other data. They found that emissions have not changed significantly since 2006, and the primary source of methane emissions is paddy fields.

Soil management in agricultural lands has a significant impact not only on sustainable crop production but also on C accumulation in the soil by influencing the soil C cycle. The relationship between soil management, soil microorganisms, and soil organic matter has been studied from various angles. Wei et al. [8] focused on the effects of cover crops in mango orchards and how they affect the soil microflora.

Liu et al. [9] focused on an autotrophic microbe that is usually overlooked in research. Their studies indicate that the bacterial community structure, composition, and CO_2-fixing capability are highly regulated by soil management practices and that minimal disturbance to the soil's microenvironment, coupled with the retention of crop residues in the soil, will improve bacteria-involved biological activities and increase nutrient cycling and soil productivity.

Li et al. [10] compared Integrated Soil-crop System Management (ISSM) based on crop models and nutrient management with conventional farming methods. They found that ISSM not only increased yields but also made soil organic C more persistent and complex within microbial networks.

Conflicts of Interest: The authors declare no conflict of interest.

References

1. Howlett, D.; Stewart, J.; Inoue, J.; Saito, M.; Lee, D.; Wang, H.; Yamada, T.; Nishiwaki, A.; Fernández, F.; Toma, Y. Source and Accumulation of Soil Carbon along Catena Toposequences over 12,000 Years in Three Semi-Natural *Miscanthus sinensis* Grasslands in Japan. *Agriculture* **2022**, *12*, 88. [CrossRef]
2. Komal, N.; Zaman, Q.; Yasin, G.; Nazir, S.; Ashraf, K.; Waqas, M.; Ahmad, M.; Batool, A.; Talib, I.; Chen, Y. Carbon Storage Potential of Agroforestry System near Brick Kilns in Irrigated Agro-Ecosystem. *Agriculture* **2022**, *12*, 295. [CrossRef]
3. Liu, C.; Li, J.; Sun, W.; Gao, Y.; Yu, Z.; Dong, Y.; Li, P. Temporal and Spatial Variations in Soil Elemental Stoichiometry Coupled with Alterations in Agricultural Land Use Types in the Taihu Lake Basin. *Agriculture* **2023**, *13*, 484. [CrossRef]
4. Matsuoka-Uno, C.; Uno, T.; Tajima, R.; Ito, T.; Saito, M. Liming and Phosphate Application Influence Soil Carbon and Nitrogen Mineralization Differently in Response to Temperature Regimes in Allophanic Andosols. *Agriculture* **2022**, *12*, 142. [CrossRef]
5. Zhu, L.; Cao, M.; Sang, C.; Li, T.; Zhang, Y.; Chang, Y.; Li, L. Trichoderma Bio-Fertilizer Decreased C Mineralization in Aggregates on the Southern North China Plain. *Agriculture* **2022**, *12*, 1001. [CrossRef]
6. Novotny, E.H.; Balieiro, F.d.C.; Auccaise, R.; Benites, V.D.M.; Coutinho, H.L.D.C. Spectroscopic Investigation on the Effects of Biochar and Soluble Phosphorus on Grass Clipping Vermicomposting. *Agriculture* **2022**, *12*, 1011. [CrossRef]
7. Wang, G.; Liu, P.; Hu, J.; Zhang, F. Spatiotemporal Patterns and Influencing Factors of Agriculture Methane Emissions in China. *Agriculture* **2022**, *12*, 1573. [CrossRef]
8. Wei, Z.; Zeng, Q.; Tan, W. Cover Cropping Impacts Soil Microbial Communities and Functions in Mango Orchards. *Agriculture* **2021**, *11*, 343. [CrossRef]

9. Liu, C.; Xie, J.; Luo, Z.; Cai, L.; Li, L. Soil Autotrophic Bacterial Community Structure and Carbon Utilization Are Regulated by Soil Disturbance—The Case of a 19-Year Field Study. *Agriculture* **2022**, *12*, 1415. [CrossRef]
10. Li, Q.; Kumar, A.; Song, Z.; Gao, Q.; Kuzyakov, Y.; Tian, J.; Zhang, F. Altered Organic Matter Chemical Functional Groups and Bacterial Community Composition Promote Crop Yield under Integrated Soil-Crop Management System. *Agriculture* **2023**, *13*, 134. [CrossRef]

Article

Temporal and Spatial Variations in Soil Elemental Stoichiometry Coupled with Alterations in Agricultural Land Use Types in the Taihu Lake Basin

Chonggang Liu [1,2,3,†], Jiangye Li [4,5,6,†], Wei Sun [1,2], Yan Gao [4,5,6,*], Zhuyun Yu [7], Yue Dong [8] and Pingxing Li [1,2,*]

1   Nanjing Institute of Geography & Limnology, Chinese Academy of Sciences, Nanjing 210008, China
2   Key Laboratory of Watershed Geographic Sciences, Chinese Academy of Sciences, Nanjing 210008, China
3   University of Chinese Academy of Sciences, Beijing 100049, China
4   Institute of Agricultural Resources and Environment, Jiangsu Academy of Agricultural Sciences, Nanjing 210014, China
5   Key Laboratory of Agricultural Environment on the Lower Yangtze River Plain, Ministry of Agriculture and Rural Affairs, Nanjing 210014, China
6   National Agricultural Experimental Station for Agricultural Environment, Luhe, Ministry of Agriculture and Rural Affairs, Nanjing 210014, China
7   Hebei Province Key Laboratory of Wetland Ecology and Conservation, Hengshui University, Hengshui 053000, China
8   Scientific Research Base Administration Office, Jiangsu Academy of Agricultural Sciences, Nanjing 210014, China
*   Correspondence: ygao@jaas.ac.cn (Y.G.); pxli@niglas.ac.cn (P.L.)
†   These authors contributed equally to this manuscript.

Citation: Liu, C.; Li, J.; Sun, W.; Gao, Y.; Yu, Z.; Dong, Y.; Li, P. Temporal and Spatial Variations in Soil Elemental Stoichiometry Coupled with Alterations in Agricultural Land Use Types in the Taihu Lake Basin. *Agriculture* **2023**, *13*, 484. https://doi.org/10.3390/agriculture13020484

Academic Editors: Yinglong Chen, Masanori Saito and Etelvino Henrique Novotny

Received: 27 December 2022
Revised: 7 February 2023
Accepted: 15 February 2023
Published: 17 February 2023

Abstract: Soil elemental stoichiometry, expressed as the ratios of carbon (C), nitrogen (N), and phosphorus (P), regulates the biogeochemical processes of elements in terrestrial ecosystems. Generally, the soil C:N:P stoichiometry characteristics of agricultural ecosystems may be different from those of natural ecosystems, with distinct temporal and spatial variations along with the alterations of agricultural land use types (LUTs). The balance of soil C, N, and P reflected by their stoichiometry is primarily important to microbial activity and sustainable agricultural development. However, information on soil stoichiometric changes after long-term alterations in land use is still lacking. We characterized the temporal and spatial changes in soil elemental stoichiometry coupled with alterations in agricultural LUTs in the Taihu Lake basin. By using the ArcGIS method and meta-data analysis, our results showed that the C:N, C:P, and N:P ratios of agricultural soil in the Taihu Lake basin were much lower than the well-constrained values based on samples from forest, shrubland, and grassland at a global scale. Generally, these elemental ratios in soils increased from the 1980s to the 2000s, after experiencing changes from agricultural to other land use. The soil C:N:P stoichiometry may have maintained the increasing trend according to the meta-data analysis from the 341 peer-reviewed publications since 2010. Nevertheless, different regions showed inconsistent change patterns, with the Tianmu Mountain area surrounding the downstream of the Taihu Lake basin experiencing a reduction in those ratios. The changes in LUTs and their corresponding management practices were the major drivers shaping the spatial and temporal distributions of soil C:N, C:P, and N:P. Paddy soil generally achieved higher C sequestration potential due to more straw input and a more rapid transfer of straw C into soil C in the upstream of the Taihu Lake basin than other land use types. These results provide valuable information for the agricultural system of intensive cultivation on how their soil elemental stoichiometry characteristics vary temporally and spatially due to the alteration of agricultural land use types.

Keywords: temporal-spatial distributions; C:N:P stoichiometry characteristics; land use type changes; sustainable agricultural development; C sequestration

1. Introduction

Life in terrestrial ecosystems is supported by many macroelements, such as carbon (C), nitrogen (N), and phosphorus (P). Their concentrations, especially their stoichiometry, not only regulate the biodiversity and growth of plants and microorganisms but also determine soil organic matter (SOM) decomposition-accumulation dynamics and C stocks, which commonly support soil health [1–3]. These provide essential ecosystem functions and services, such as the promotion of sustainable food production, maintenance of soil fertility, regulation of greenhouse gas emissions, and regulation of water quality in watersheds [4]. Therefore, the soil elemental stoichiometry (e.g., C:N:P), which considers the balance of energy (C) and nutrient elements (N, P), has attracted increasing attention.

Soil total organic carbon (TOC) dynamics are regulated by elemental stoichiometry [5,6]. As the dominant constituent of SOM, TOC represents one of the most commonly determined soil attributes. The TOC/TN (total N) ratio (C:N) provides valuable information concerning the status of decomposition and potential nutrient availability in organic matter [7,8]. TN/TP (Total P) and TOC/TP have also been widely recognized and studied because N and P are the main growth-limiting nutrients for plants and microorganisms in soil [9]. The stoichiometry between C, N, and P (C:N, C:P, N:P) might regulate SOM dynamics and soil microbial carbon use efficiency (CUE) [10–13]. Currently, most studies mainly focus on the effect of different fertilization treatments on soil stoichiometry; the effects of land cover and land use, however, have rarely been mentioned [14].

Alterations in land cover and land use, driven by anthropogenic activities, are a global phenomenon that may affect the soil elemental stoichiometry [15,16]. Since the last century, the growing demand for food has led to intensive agriculture in China due to the surge in populations, and large numbers of forests or grasslands have been replaced by agricultural land. Vegetation have been drastically modified as a result, and soils have been exposed to radical changes, mainly in plant diversity, water balance, and C and nutrient dynamics [17–20]. Moreover, agronomic practices involve many mechanical and chemical disturbances that modify soil chemistry, physics, and its microbial communities [21–23]. It is well documented that land-use changes for various agricultural purposes have resulted in losses of 25~50% of soil organic carbon (SOC) in soil originating from forests [24,25], shrublands [26], and grasslands [27]. Moreover, modern agriculture, with its imbalance in the use of N and P fertilizer, may create an unprecedented human-induced imbalance of C:N:P ratios in agricultural soil [28,29]. The combination of these factors could generate dramatic stoichiometric shifts in soils due to land-use change (LUC). In addition, different crops vary in the demand for fertilizer and management practices, which are bound to induce the disparity of soil C:N:P stoichiometry and their C storage capacities between different croplands [30,31]. However, the temporal and spatial changes in elemental stoichiometry in agricultural soil and their relations with LUC remain unknown.

The Taihu Lake basin provides an excellent opportunity to study the temporal and spatial changes in elemental stoichiometry in agricultural soil and their relations with LUT changes due to its unique environmental and climatic characteristics and LUT changes resulted from fast-growing socioeconomic conditions in China. The land surface in this region undergoes annual disturbance and change through urbanization, logging of forestry to grow trees of economic value, changing cropland to orchard or vegetable land, and farming practices such as crop rotation and ploughing [32,33]. This region has been regarded as one of the major grain-producing areas in China. Its warm weather and sufficient water supply allow the growth of two crops a year, e.g., double-cropping of rice or rice–wheat rotation. However, the agricultural land use types in this region are experiencing unprecedented change due to the aim of higher profits. In 1985, the rice and wheat planting area in this region was approximately 171.8×10^4 hm^2 [34,35], while the main crop planting area (115.5×10^4 hm^2) was reduced by 32.8% in 2010 [36]. Much SOC is at risk of being transferred to the atmosphere via deforestation, land clearing, and other forms of land use type alteration driven largely by an increasing population [36,37]. Moreover, the unbalanced use of N and P fertilizers may aggravate the imbalance of soil C:N:P stoichiometry. As a result, this may lead to alterations in ecosystem function and services.

In this study, our major objectives were (1) to characterize the temporal and spatial changes in topsoil C, N, and P concentrations and their ratios, and (2) to further investigate the correlations between the concentration of C, N, or P and the corresponding C:N or C:P to reflect whether the ecological stoichiometry were still able to regulate the stocks of soil C, N, or P, with alterations in land use types in the Taihu Lake basin from the 1980s to the 2010s. In the present study, one hypothesis was tested: the soil C:N, C:P, and N:P ratios increased with time, especially in paddy fields. This research is critically important to learn how land-use changes impact ecosystem services, and would provide practical significance to proper land use for sustainable agricultural development.

2. Material and Methods

2.1. Study Area

The Taihu Lake basin (Figure 1A), located on the east coast of China (119°3′1″~121°54′26″ E, 30°7′19″~32°14′56″ N), is one of the most developed regions of China and it was where the land use types experienced a dramatic change from 1980 to 2016 with the development of the economy (Figure 1B). Taihu Lake (2238.1 hm^2) is the largest lake in this region and the third largest freshwater lake in China. The basin has a typical subtropical monsoon climate, with an annual mean temperature of 15~17 °C and annual mean precipitation of 1010~4000 mm. The dominant soil types were yellow brown soil and paddy soil.

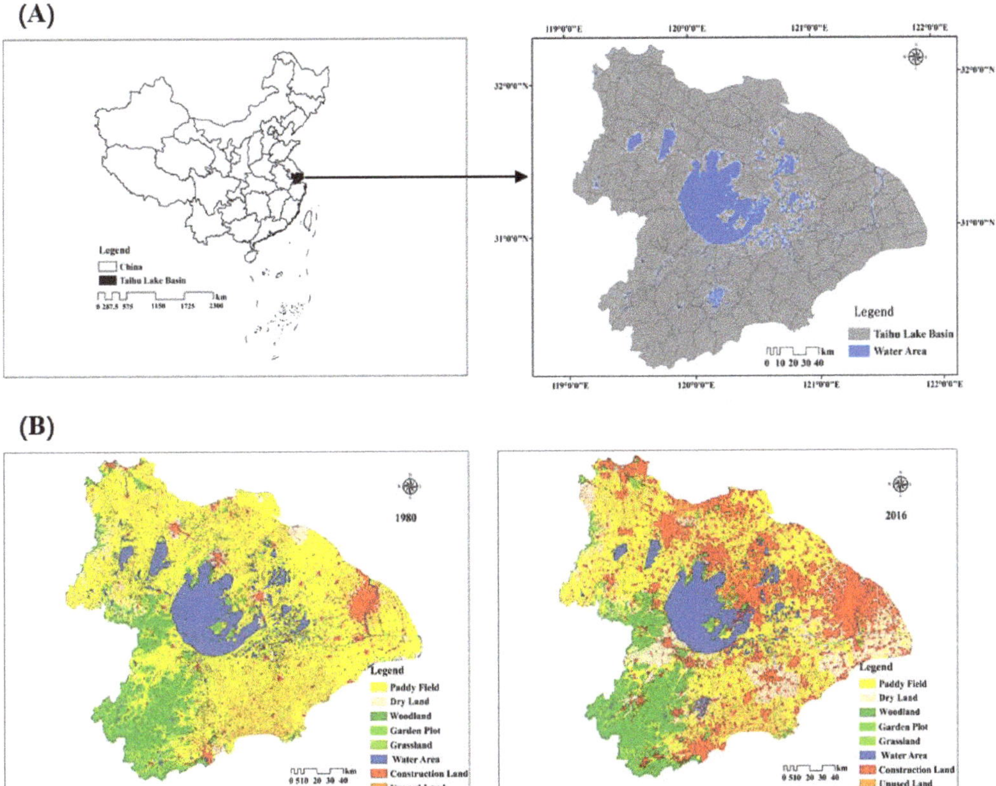

Figure 1. The site of the Taihu Lake basin (**A**) and land use type changes from 1980s to 2000s in Taihu Lake basin (**B**).

Generally, there was a net increase in woodland area for these two periods (11.42 hm^2 from 1980 to 2000; 137.72 hm^2 from 2000 to 2016), which was mainly converted from dry land during 1980–2000 and from paddy fields, dry land, and grassland during 2000–2016 (Table S1). However, there was a net decrease in grassland area for these two periods (3.8 hm^2 from 1980 to 2000; 81.18 hm^2 from 2000 to 2016), which was mainly converted to construction land during 1980–2000 which was then converted to woodland during 2000–2016.

2.2. Data Collection

We collected soil TN, TP, and SOC data for Taihu Lake basin in 1980 from the second Chinese Soil Survey, and the data for the year 2000 from Soil Series Survey for Jiangsu and Zhejiang Province. These two surveys used the same sampling sites. There was a total of 1451 data for each parameter of TN, TP, and SOC in each survey. In this study, only the surface (0 to 20 cm) soil data were used as it directly recorded the impact of land-use changes in the time scale of decades. We collected the data after 2010 from peer-reviewed papers published in both Chinese and English journals with regard to major ALUTs in Taihu Lake region, e.g., one season rice, rice–wheat rotation, orchard, tea plantation. Studies in Chinese were collected from the China National Knowledge Infrastructure (CNKI) database and those in English from ISI-Web of Science. The key words used for searching were soil organic matter, soil organic carbon, total nitrogen, total phosphorus, paddy field, wheat, orchard (e.g., grape, peach, pear), tea plantation. Papers published were only included if they met the following criteria: (1) studies had to report at least two variables including soil organic matter/soil organic carbon, total nitrogen, total phosphorus in order to calculate C:N, C:P, and N:P stoichiometry; (2) any studies lacking replication were not considered; (3) only the experimental data obtained between 2015 and 2020 were considered to represent more recent soil characteristics. A total of 341 peer-reviewed publications (204 observations for paddy field, 62 observations for rice–wheat rotation, 46 observations for Orchard, 29 observations for tea plantation) were selected for our analysis.

Among them, a total of 108 samples were measured in laboratory in 2018. TOC, TN, and TP contents were determined by a CHNS elemental analyser (Carlo Erba model EA1108, Italy Vario). The total P (TP) content in the soil was measured using perchloric acid digestion [38]. Additionally, the soil stoichiometry was calculated by the ratio of TOC to TN (C:N), TOC to TP (C:P), and TN to TP (N:P).

2.3. Data Processing and Calculation

2.3.1. Mapping Method for Spatial Distribution and Variation of Soil C:N, C:P, and N:P in 1980~2000

Based on the 1451 data that we obtained from China's second soil survey (1980) and the Soil Series Survey for Jiangsu and Zhejiang Province in 2000, the 1451 sampling points were generated according to the latitude and longitude in ArcGIS 10.8. The mass ratio of C:N, C:P, and N:P was calculated based on the concentrations of soil TOC, TP, and TN. The changing rates for these parameters between 1980 and 2000 were calculated according to their differences in 2000 and 1980. The Kriging method was used to interpolate and obtain the spatial distribution of C:N, C:P, N:P, and their changing rates in the Taihu Lake basin, respectively. The Kriging method weighed the area between multiple sampling points to derive an estimate of the unmeasured position. This method is similar to the inverse distance weighting method, which consists of a weighted sum of the data:

$$\hat{Z}(s_0) = \sum_{i=1}^{N} \lambda_i Z(s_i) \tag{1}$$

where $Z(s_i)$ represents the measured value at the position of the i-th sample point, λ_i represents the unknown weight of the measured value at the position of the i-th sample point, and the weight λ_i depends on the distance between the sample point, the predicted position, and the predicted position A fitted model of the spatial relationship between the measured values of the sample points; s_0 represents the predicted position; N represents the measured value of the sampled point.

The "nature break" method was selected to cluster the spatial interpolation. These classes were based on natural groupings inherent in the data. ArcMap 10.8 identified break points by picking the class breaks that best grouped similar values and maximized the differences between classes. The features were divided into classes whose boundaries were set where there were relatively big jumps in the data values.

2.3.2. Computation Method of Proportion of Different Land Use Types in Each Index Interval

The changes of C:N, C:P, and N:P between 1980 and 2000 were intersected with land-use changes in 1980 and 2000, respectively. The intersected layers were recalculated while different land use types and their areas in 1980 and 2000 were extracted according to the divided intervals as described in Section 2.3.1. Through the comparison of the land use data between the two periods, we could obtain the main land use type changes in different C:N, C:P, and N:P changing intervals, and then analyze the correlations between them. These classes were based on natural groupings inherent in the data. ArcMap identified break points by picking the class breaks that best grouped similar values and maximized the differences between classes. The features were divided into classes whose boundaries were set where there were relatively big jumps in the data values.

2.3.3. Statistical Analysis

The ratios of soil C:N, C:P, and N:P in 1980 and 2000 in the Taihu Lake basin were also shown in box plots conducted by Origin 9.1. The differences of C:N, C:P, and N:P between 1980 and 2000 were measured by paired-sample *t*-test in SPSS 20.0 with a 95% confidence interval. Linear regression analysis was conducted after the Pearson product-moment correlation analysis by two-tailed test in SPSS 20.0. Correlations between topsoil C:N, C:P, or N:P and corresponding C, N, or P concentrations were conducted in Excel 2016 and the statistical analysis was conducted using a *t*-test in SPSS 20.0.

3. Results

3.1. Temporal and Spacial Changes in Soil C:N:P Stoichiometry in the Taihu Lake Basin

In the 1980s (Figure 2A), the soil C:N in most areas of the Taihu Lake basin ranged from 9 to 10 and 10 to eleven, accounting for 42.6% and 26.5% of the total area, respectively. A total of 19.8% of the area showed a soil C:N ratio lower than 9, while only 11.1% of the area showed a soil C:N ratio higher than 11. In the 2000s (Figure 2B), the area with soil C:N ranging from 10 to 11 greatly increased when compared to the 1980s, with 42.4% of the area showing soil C:N ranging from 10 to 11. Only 4.76% of the area presented a soil C:N ratio lower than 9, while 15.61% of the area showed a C:N ratio higher than 11.

In the 1980s (Figure 2C), the soil with C:P lower than 28 accounted for 61.88% of the total area (20.9% of the total area had C:P < 18; 27.3% of the total area had C:P ranging from 18 to 24; 12.4% of the total area had C:P ranging from 24 to 28). In the 2000s (Figure 2D), the soil C:P ratio in the Taihu Lake basin was generally enhanced, with the area of soil C:P > 28 accounting for 67.0% of the total area.

In the 1980s (Figure 2E), the soil N:P ratio in the Taihu Lake basin was mainly distributed in the ranges of ≤ 2, two~three, and 2~4, which accounted for 24.7%, 38.1%, and 28.0% of the total area, respectively. Soil with N:P ≥ 4 accounted for only 9.3% of the total area in the Taihu Lake basin. In the 2000s (Figure 2F), the area with soil N:P ranging from 2 to 3, 3 to 4 and ≥ 4 increased to 46.0%, 30.9%, and 13.1% of the total area, respectively, while the area with soil N:P ≤ 2 decreased to 10.1% of the total area.

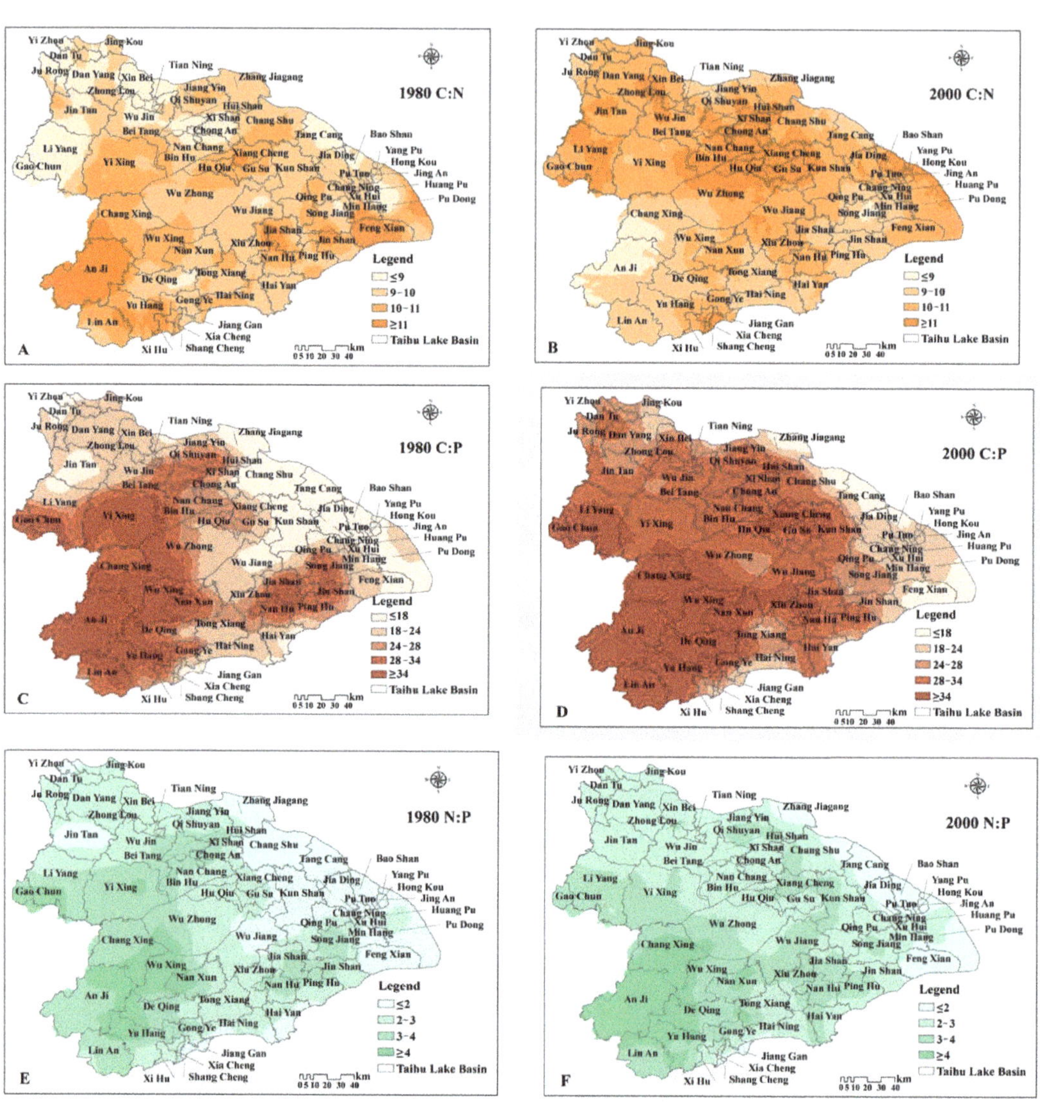

Figure 2. Temporal and spatial distribution of topsoil C:N, C:P, and N:P in 1980s (**A**,**C**,**E**) and 2000s (**B**,**D**,**F**), respectively, in Taihu Lake basin.

According to Figure 2 and relevant calculations, in the 1980s, the soil C:N ranged from 9 to 10 was mainly from paddy fields (87.8%), while the soil C:N ranged from 10 to 11 was from both some paddy fields (59.2%) and woodlands (32.2%). Most of these areas were evenly distributed in the central Taihu Lake basin. A total of 19.8% of the area showed a soil C:N ratio lower than 9 and was mainly scattered in the northwest Taihu Lake basin. Only 11.08% of the area showed a soil C:N ratio higher than 11 and was mainly scattered in the region surrounding the southwestern Taihu Lake basin, especially in the mountainous area (44.8%) in Anji and Lin'an of Zhejiang Province.

In the 2000s, the area with soil C:N ratios ranging from 10~11 was mainly distributed in paddy fields (85.7%) between Nanjing and Shanghai located in the northern Taihu Lake basin. A total of 15.6% of the total area showed C:N higher than 11 distributed in Wuxi and

Changzhou, where both industry and agriculture are well developed. In contrast, 4.8% of the area with soil C:N lower than nine was mainly distributed in the mountainous area of Anji and Lin'an of Zhejiang Province, which used to have high soil C:N in the 1980s.

In the 1980s, the soil with a C:P ratio lower than 28 was mainly distributed in paddy fields (86.9~88.8%) located in the region surrounding the downstream Taihu Lake basin along the Yangtze River. The soil with a C:P higher than 28 was mainly distributed in woodland areas and paddy fields located in the mountainous regions surrounding the upstream of the Taihu Lake basin.

In the 2000s, the soil C:P ratio in the Taihu Lake basin was generally enhanced, especially in the region surrounding the downstream of the Taihu Lake basin along the Yangtze River, in some areas surrounding large- and medium-sized cities (e.g., Suzhou and Wuxi), which used to have low C:P in the 1980s. The area with soil C:P ranging from 28 to 34 was mainly distributed in the paddy field (69.8%), while soil with C:P > 34 was mainly distributed in both the mountainous area (41.1%) and some paddy fields (50.9%).

In the 1980s, relatively high soil N:P, e.g., N:P ranged from three to four and ≥four, was mainly distributed in the woodland area and some paddy fields located in the region surrounding the downstream Taihu Lake basin along the Yangtze River. Relatively lower soil N:P, e.g., N:P ranged from 2 to 3 and ≥2, was mainly distributed in some paddy fields located in the mountainous regions surrounding the upstream Taihu Lake basin. In the 2000s, higher soil N:P was also mainly distributed in woodlands and some paddy fields, while lower soil N:P was mainly distributed in some paddy fields.

The box plots showed that the median values of soil C:N, C:P, and N:P in the 2000s were consistently higher than those observed in the 1980s (Figure 3, $p < 0.05$). The mean C:N, C:P, and N:P mass ratios of the surface layer (0~20 cm) soil in the Taihu Lake basin were 9.6, 21.9, and 2.3 in the 1980s and 10.4, 30.1, and 2.8 in the 2000s, respectively, and the differences in the same elemental stoichiometry between the 1980s and 2000s were all significant ($p < 0.05$).

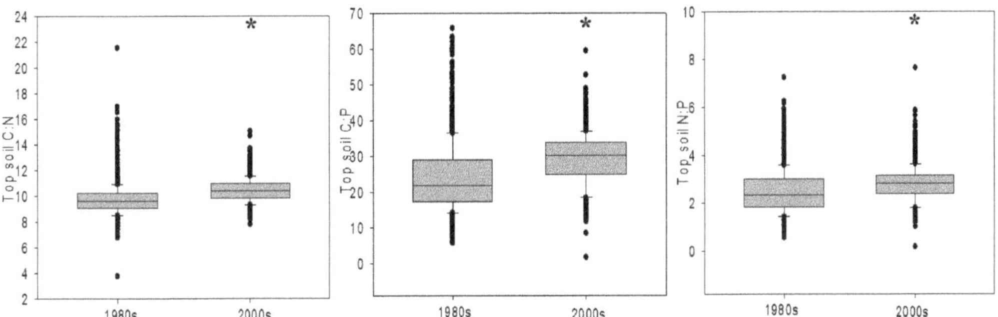

Figure 3. The vertical box plot of topsoil C:N, C:P, and N:P ratios in 1980s and 2000s, respectively, in Taihu Lake basin. The boundary of the box closest to zero indicated the 25th percentile, a solid line within the box marked the median, and the boundary of the box farthest from zero indicated the 75th percentile. Whiskers (error bars) above and below the box indicated the 90th and 10th percentiles. "*" indicates the difference was significant at 5% level.

3.2. Alteration of Soil C:N, C:P and N:P Coupled with Agricultural Land Use Type Changes between 1980 and 2000

From the 1980s to the 2000s, paddy fields suffered dramatic changes, accounting for 80.8% of the total altered agricultural land area (the sum of paddy field, dryland, woodland, grassland, and unused land). Nevertheless, paddy fields were still the major LUTs in this region, consisting of 75.5% of the land area with the LUTs remained unchanged (Table 1). Among the unchanged paddy fields, 69.0% of the fields experienced a substantial increase in soil C:N (21.0% of paddy had an increase in C:N < 10%; 24.9% of paddy fields had an increase in C:N ranging from 10 to 20%; 15.9% of paddy fields had an increase in C:N > 20%), whereas 31.0% of paddy fields experienced an obvious reduction in soil C:N (21.6% of paddy fields had a decrease in C:N < 10%; 9.3% of paddy fields had a decrease in C:N > 10%). Although the paddy fields were mainly occupied by construction land, they were transformed to drylands and woodlands. Most of these agricultural soils (85.2%) alternated from paddy fields showed an obvious increase in soil C:N (25.1% of paddy fields had an increase in C:N <10%; 36.2% of paddy fields had an increase in C:N ranging from 10 to 20%; 23.9% of paddy fields had an increase in C:N >20%), whereas 14.8% of paddy fields experienced an obvious reduction in soil C:N (13.4% of paddy fields had a decrease in C:N < 10%; 1.3% of paddy fields had a decrease in C:N > 10%).

Table 1. Alteration of soil C:N, C:P, and N:P coupled with agricultural land use types changes between 1980 and 2000.

		Unchanged (hm^2)					Changed (hm^2)				
		Paddy Field	Dry Land	Woodland	Grassland	Unused Land	Paddy Field	Upland Field	Woodland	Grassland	Unused Land
Changes of soil C: N (%).	≤−20	77,853.79	10,942.83	147,429.43	5874.96	16.65	620.08	1656.62	428.58	0.00	12.43
	−20−−10	142,694.02	9422.41	108,912.23	1410.83	35.32	1796.95	1526.09	206.57	0.00	4.04
	−10–0	513,178.59	22,414.50	164,748.59	4850.84	166.26	23,873.02	5031.70	983.25	374.41	30.41
	0–10	668,724.27	49,357.48	80,318.71	2249.48	425.65	44,534.01	8460.33	1794.84	146.88	69.92
	10–20	590,317.97	38,555.91	35,746.25	1130.68	623.65	64,177.17	9330.64	1363.75	99.41	157.47
	≥20	377,667.73	57,929.54	24,923.92	1208.59	210.26	42,502.03	10,185.77	279.68	4.59	0.00
Changes of soil C: P (%)	≤−20	146,587.00	9108.14	26,910.83	1097.35	128.76	6902.43	1171.05	294.70	0.00	17.67
	−20–0	438,790.96	30,831.96	196,993.76	3966.87	27.58	17,470.28	4686.35	656.82	368.63	20.93
	0–20	528,502.10	40,372.40	199,546.41	6205.09	744.01	38,351.01	8139.33	1192.67	101.40	42.95
	20–40	494,907.77	46,879.72	107,773.23	4412.91	341.55	53,162.51	11,567.71	2013.63	124.78	0.00
	40–60	270,568.22	22,323.22	20,865.31	544.00	183.68	21,384.68	5624.71	581.01	5.29	157.43
	≥60	491,080.22	39,106.73	9989.76	499.14	52.19	40,232.36	5001.80	317.86	25.18	35.30
Changes of soil N: P (%)	≤−20	81,472.90	9884.22	17,319.89	1310.53	131.83	5881.10	1017.75	101.06	11.12	23.29
	−20–0	594,164.88	44,217.47	83,620.59	2435.38	257.54	39,964.58	8941.48	708.17	145.79	8.82
	0–20	670,129.27	57,037.72	214,093.68	3548.60	616.37	60,134.29	12,003.25	1742.88	313.09	26.30
	20–40	459,801.07	37,325.82	108,106.22	3976.50	354.78	32,614.62	7035.71	1818.24	130.10	174.64
	40–60	228,020.72	20,451.19	55,808.84	2713.29	100.61	9792.92	2368.16	318.42	0.00	0.00
	≥60	336,847.79	19,706.46	83,129.97	2741.06	16.65	29,115.76	4824.60	367.90	25.18	41.24

Except for the construction land, a large area of dry land was transformed to paddy fields and woodlands. A total of 77.3% of the altered dry land had a dramatic increase in soil C:N, e.g., 23.4% of dry land showed an increase of <10%; 25.8% of dry land showed an increase in C:N ranging from 10 to 20%; and 28.1% of dry land showed an increase in C:N > 20%. Meanwhile, 22.7% of the alternating dry land experienced a decrease in soil C:N. The dry land that remained unchanged had a similar increasing trend for soil C:N, with 77.3% of the unchanged dry land showing a substantial increase in soil C:N, but 22.7% of the unchanged dry land showing an obvious decrease in soil C:N.

The woodland area generally experienced little change. However, there was a contrasting trend for soil C:N changes between woodland areas that underwent alteration and remained unchanged. Among woodlands remaining unchanged, 74.9% of woodlands soil experienced an obvious decrease in soil C:N, with 29.3% woodland area showing a decrease in C:N < 10%; 19.4% woodland area showing a decrease in C:N ranging from 10 to 20%; and 26.2% woodland area showing a decrease in C:N > 20%. In contrast, among woodlands that has been transformed to other LUTs, 68.0% of woodland soil had an obvious increase in soil C:N, with 35.5% woodland area showing an increase in C:N < 10%; 27.0% woodland area showing an increase in C:N ranging from 10 to 20%; and 5.5% woodland area showing an increase in C:N > 20%.

Most paddy fields, drylands and woodlands, regardless of their LUT changes, experienced increases in soil C:P and N:P. A total of 75.3% and 71.5% of the paddy fields that remained unchanged increased in soil C:P and N:P, respectively. Meanwhile, 86.3% and 74.2% of the altered paddy field experienced a substantial increase in soil C:P and N:P as well, respectively. In addition, 78.8% and 71.3% of the dry land that remained unchanged had increases in soil C:P and N:P, respectively. Similarly, 83.8% and 72.5% of the alternated dry land experienced a substantial increase in soil C:P and N:P, respectively.

3.3. Responses of Soil Elemental Stoichiometry to the Changes in the C, N, and P Pools in Soil between the 1980s and 2000s

In the 1980s, topsoil C:N was significantly positively correlated with soil TOC, whereas it had no significant correlation with soil TN (Figure 4). In contrast, in the 2000s, topsoil C:N did not show a significant correlation with soil TOC, but it was significantly negatively correlated with soil TN. The topsoil C:P in the 1980s and 2000s was significantly positively correlated with soil TOC, while it was significantly negatively correlated with soil TP. The topsoil N:P in the 1980s and 2000s were all negatively correlated with soil TP, while they showed a positive correlation with soil TN in the 1980s but a negative correlation with soil TN in the 2000s. Extremely similar correlation patterns between topsoil N:P or C:P and TP were observed for both the 1980s and 2000s, showing a decrease in N:P or C:P accompanied by an increase in soil TP concentrations.

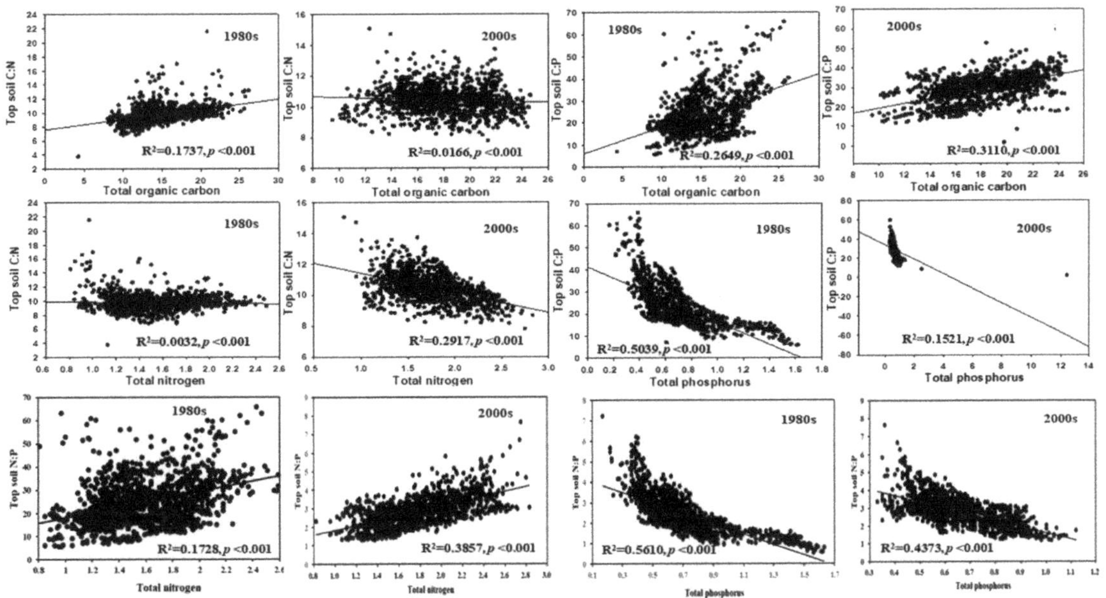

Figure 4. Correlations between topsoil C:N, C:P, or N:P and corresponding C, N, or P concentrations.

3.4. Meta-Analysis of Soil C:N, C:P, and N:P under Different Land Use Types after 2000s

The topsoil C:N, C:P, and N:P ratios under the major ALUTs (paddy fields, paddy–wheat rotation, orchard) in the Taihu Lake basin during different periods (1980s, 2000s, 2015–2020) are shown in Table 2 according to data collected from published literature. Overall, the median values of C:P for rice fields, fields with rice–wheat rotations, and orchards increased from the 1980s and 2000s to the present (2015~2018), which was consistent with the data obtained from the soil survey, as shown in the previous section. The median values of C:N for rice fields also showed an increasing trend from the 1980s to the present (2015~2018). The median values of C:P derived from the published papers were higher than the values obtained from the soil survey for the 1980s and 2000s.

Table 2. Topsoil C:N, C:P, and N:P ratio under different agricultural land use types (ALUTs) in Taihu Lake basin according to meta-data analysis.

Land Use Type	Data Number	C:N			C:P			N:P		
		Mean	Mid-Value	Min-Max	Mean	Mid-Value	Min-Max	Mean	Mid-Value	Min-Max
1980s										
Rice-Field	25	9	10	5–11	57	49	14–106	3	4	1~6
Rice–Wheat Rotation	6	12	13	10–13	41	40	34–48	3	3	2~3
Orchard	3	14	14	13–17	55	54	48–63	4	4	3~4
Tea Plantation	4	5	5	5–6	24	23	17–31	5	5	3~6
2000s										
Rice-Field	97	11	10	6–35	53	48	13–103	5	5	1~13
Rice–Wheat Rotation	26	11	10	9–19	31	26	13–49	3	3	1~5
Orchard	17	15	15	8–26	45	48	17–59	3	3	1~6
Tea Plantation	10	11	11	10–13	38	31	19–94	3	3	2~7
2015–2020										
Rice-Field	26	13	10	6–38	50	43	13–101	5	4	5~12
Rice–Wheat Rotation	14	10	10	6–13	41	42	22–56	4	4	3~6
Orchard	11	12	9	9–23	25	23	7–60	2	2	1~4
Tea Plantation	6	5	5	4–7	45	45	19–57	8	8	4~13

4. Discussion

4.1. The General Soil C:N:P Stoichiometry Characteristics in Taihu Lake Basin

The stoichiometry of C, N, and P is used to reveal the transitions in nutrient limitations for microorganisms and plants. Moreover, soil C:N:P stoichiometry characteristics also reflect the soil health condition. The soil C:N ratio has been widely recognized as a good indicator of the degree of SOC decomposition and accumulation processes. A previous study showed lower and less variable mass C:N ratios ranging from 9.8 to 12.4 in the 0~20 cm depth, compared with our results which showed that the mass C:N mean value was 9.7 in the 1980s; the C:N mean value was 10.7 in the 2000s; C:N mean value was 13 for paddy field and 10 for the field with rice–wheat rotation from 2015 to 2020 in the Taihu Lake basin [39].

In a recent global meta-data analysis, Cleveland and Liptzin [10] stated a remarkably constrained soil C:N:P ratio of 186:13:1 (molar ratio) on the global scale. This corresponded to soil C:N, C:P, and N:P molar ratios of approximately 14.3, 186, and 13, respectively. However, their analysis was mainly based on samples from forest, shrubland, and grassland. Our results revealed that the soil mean C:N, C:P, and N:P molar ratios in the Taihu Lake basin were 11.4, 61.8, and 5.4 in the 1980s and 12.5, 74.7, and 6.0 in the 2000s, respectively. Generally, the soil mean molar C:N ratio in the Taihu Lake basin was slightly lower than the above constrained values, while the soil mean C:P and N:P molar ratios were far lower than the constrained values. The major reason for these differences was probably that a large amount of C was removed after harvest in the agricultural field, causing a reduction in plant C input such as above litter, rhizodeposits, and root mass [40]. Afterwards, a nationwide meta-analysis for soil elemental stoichiometry across various soil groups in China was conducted by Tian et al. based on the second Chinese soil survey (National Soil Survey Office 1993, 1994a,b, 1995a,b, 1996) [40]. From the frequency distribution of soil C, N, and P ratios, they found that most C:N, C:P, and N:P ratios were in the range of 6~12, 24~48, and 3~6, respectively. The number-weighted mean soil C:N, C:P, and N:P ratios were 11.9, 61, and 5.2, respectively, which were similar to the area-weighted means (12.1, 61, and 5.0, respectively). Overall, our C:N:P stoichiometry values fell in the above ranges. Tian et al. [41] also revealed that the C:N, C:P, and N:P molar ratios of the surface organic-rich layer (0 cm~10 cm of A horizon) in China were 14.4, 136.0, and 9.3, respectively. This finding was similar to the well-constrained values observed by Cleveland and Liptzin [10]. However, even in the mountainous area of the Taihu Lake basin, which was believed to have high C:N ratios, the soil mean C:N, C:P, and N:P ratios were lower than the well-constrained values (soil C:N:P ratio of 186:13:1). Typically, a C:N ratio above 12~14 is considered indicative of a shortage of nitrogen in the soil. The C:N ratios between nine and twelve reflect a lower degree of decomposition of the organic materials present in the Taihu Lake basin [42].

4.2. Temporal Changes in Soil Elemental Stoichiometry in the Taihu Lake Basin

From the 1980s to the 2000s, the soil C:N, C:P, and N:P ratios generally experienced consistent increases in the Taihu Lake basin, although few specific areas experienced obvious reductions in these ratios (Figure 2). The increase in the soil C:P and N:P ratios was supposed to be due to a more dramatic increase in soil TOC by 18.9 and TN by 13.5% with slow-growing TP by 7.5% from the 1980s to the 2000s in the Taihu Lake basin (Figure S1). According to the Statistics Yearbook of China, the amount of P fertilizer consumption in China increased by approximately 1.5 times from 2.73 million tons in 1980 to 6.91 million tons in 2000, N fertilizer consumption increased by 1.3 times from 9.34 million tons in 1980 to 21.82 million tons in 2000, while compound fertilizer with a ratio of $N:P_2O_5$ ranging from two to five increased by 31 times from 0.27 million tons in 1980 to 8.64 million tons in 2000, which may have induced the increases in both soil TN and TP content, and a greater increase in TN content than TP [43]. N addition could reduce C mineralization rates [44], which indicated the increase in N fertilizer application rate in favor of soil C storage. Although there were major missing N or P fertilizer consumption data for Jiangsu and Zhejiang Provinces in the Taihu Lake basin before 1990, relevant research revealed that the consumption of N fertilizers was higher than that of P fertilizers during 1980 and 2000 [45]. In addition, the correlation analysis revealed that soil C:P ratios were regulated by both soil TOC and TP. On the one hand, the increase in TOC in the soil was supposed to be closely related to the straw return practice. Since 1990, the Chinese government has forced the return of crop straw to the field nationwide. From 1980 to 2000, the yields of rice, wheat, and corn increased approximately 34%, 80%, and 69%, respectively [34]. Increasing primary productivity involves the increased input of organic C into the soil, especially in soils with rice–wheat rotations in southern China [43,45,46]. It was reported that the C sequestration potential of paddy soils had reached 10,148 Tg C since straw return was implemented in subtropical areas of China [34,47]. Meanwhile, organic fertilizer substituting for partial chemical fertilizers and other agronomic measures that are beneficial for C sequestration were encouraged in that time [48,49], which would further enhance soil C sequestration. Moreover, chemical fertilizer application rates have been required to be reduced in the most recent 10 years and will be kept in the reduced rate in the future. This resulted in the enhanced soil C:N (10.34 in 2000s compared with 9.79 in 1980s) and C:P ratios (29.12 in 2000s compared with 24.38 in 1980s).

Similarly, the increase in soil C:N ratios was believed to be due to a more dramatic increase in soil TOC relative to a slower increase in TN. In the 1980s, the soil C:N ratios were mainly regulated by the decomposition of soil organic matter, indicating significant positive correlations with soil TOC content. However, in the 2000s, such a correlation disappeared when the soil C:N ratios in the Taihu Lake basin experienced a slight increase as a whole. This may indicate decoupled cycles of soil C and N in the Taihu Lake basin due to the exogenous input of inorganic N fertilizers as well as organic materials, e.g., straw. According to the meta-data analysis from the published paper, the soil C:N and N:P ratios may maintain an increasing trend in most regions of the Taihu Lake basin [41].

4.3. Spatial Changes of Soil Elemental Stoichiometry in Taihu Lake Basin

The Taihu Lake basin presented obvious spatial differences in soil C:N:P stoichiometry characteristics. Generally, the region surrounding the downstream Taihu Lake basin experienced a great reduction in soil C:N, C:P, and N:P between the 1980s and 2000s, especially the Tianmu Mountain area located in northeast Zhejiang Province, e.g., Anji and Lin'an city. The sharp reduction in soil C:N mainly resulted from the dramatic increase in exogenous N input between 1980 and 2000 due to the development of economic forests (e.g., walnut trees, Chinese chestnut trees, tea-oil trees, and bamboo forests) and tea plantations in those regions. Although there was a major lack of statistical data for the tea plantation area before 2010, the data thereafter clearly showed an increase in the tea plantation area in Zhejiang Province from 13,567 ha in 2011 to 198,524 ha in 2017. Such enhanced forestry output implied the combination of the increasing economic forest area and the production that is accompanied by increased N fertilizer input. The N fertilizer amount used in northeastern Zhejiang Province (total is 24.81 tons in 2017), where Anji, Lin'an, and Jiaxing (Tianmu Mountain area) are located, was higher than that in southwestern Zhejiang Province (the total was 18.09 tons in 2017). At the same time, this region (e.g., surrounding Lin'an city) experienced a slight decrease in soil TOC content. As a result, the soil C:N ratios in the region underwent a dramatic decrease. Similarly, the decrease in soil C:P in some regions of the Tianmu Mountain area was also mainly due to the increase in P input but a slight decrease in soil TOC. The increase in soil TN was greater than that in soil TP, resulting in the increase in soil N:P at the same time.

Moreover, the regions with fast economic growth, e.g., Kunshan, Wujiang, Changsu, and Tangcang, upstream of the Taihu Lake basin generally experienced an increase in soil C:N, C:P, and N:P between 1980 and 2000. The obvious increase in C:P in these regions was mainly due to a sharp decrease in soil P (Figure S2). Moreover, the increase in C:N in these regions mainly resulted from the decrease in soil N. The extent of the decrease in soil TP was larger than that of soil TN, resulting in the increase in soil N:P. In these developed areas with the rapid development of small- and medium-scale factories and tourist agriculture, the traditional agriculture that required large exogenous N and P inputs was reduced. As a consequence, soil C:N, C:P, and N:P experienced obvious increases.

As the spatial distribution of soil elemental stoichiometry was closely correlated with LUTs in different regions of the Taihu Lake basin, the changes in LUTs in these regions in the future will further shape the spatial distribution pattern.

4.4. The Effect of Agricultural Land Use Types on the Variation in Soil Elemental Stoichiometry

In the Taihu Lake basin, the variations in soil C:N, C:P, and N:P were mainly related to ALUTs and their corresponding management methods. Although a large area of agricultural land was transformed to construction land, the national soil surveys for soil C, N, and P were derived from agricultural lands, e.g., paddy fields, drylands, grassland, and woodlands. Paddy fields were the major ALUTs in this region, most of which experienced a substantial increase in soil C:N from the 1980s to the 2000s. Meanwhile, some paddy fields underwent a decrease in soil C:N. Normally, mechanical tillage was widely employed, which could accelerate SOC and N mineralization of agricultural land [50], thereby lowering C:N ratios as C losses as CO_2 have been expectedly higher than those for N. This trend can be even stronger as N fertilization or biological fixation adds more N to the system, which can also enhance SOC depletion by priming effects [51]. Similarly, low C:N ratios have been reported for soils under leguminous crops where N fixation is efficient and SOC decomposition is accelerated or in acidic soils [52]. However, an increase in the soil C:N of agricultural soil can occur when the supply of organic material (e.g., straw) substantially exceeds the increase in N fertilizer use. Similarly, most dry land area saw an increasing trend for soil C:N.

The changes in paddy fields and dry land were mainly transformed to each other or woodlands. Most of the transformed fields underwent an obvious increase in soil C:N, whereas some of them had a substantial decrease in soil C:N. This means that the agricultural soils generally experienced an increase in soil C:N after the rotation pattern changed. We speculate that the decrease in soil C:N was mainly from the alteration of paddy fields or drylands to economic forests that generally received excessive exogenous N input but without the simultaneous input of a large quantity of organic material (e.g., straw), as we discussed in the previous section [39]. The total woodland area experienced little change during 1980–2000. However, there was a contrasting trend for soil C:N changes between woodland areas that underwent alteration and remained unchanged. Among woodlands that had been transformed to other ALUTs, 68.0% of woodland soil had an obvious increase in soil C:N, as they were mainly transformed to paddy fields and dry land. In addition, most paddy fields, drylands soils, and woodlands, regardless of their LUT changes, experienced an increase in soil C:P and N:P. This is mainly because the exogenous input of organic materials and N fertilizer contributed to the increase in soil C:P and N:P, which exceeded the increase in P input.

5. Conclusions

The C:P (11:1~59:1) and N:P (1.2:1~7.6:1) ratios of agricultural soil in the Taihu Lake basin were far lower than the well-constrained values (C:N:P = 186:13:1) based on samples from forest, shrubland, and grassland at a global scale. The soil C:N was slightly lower than the well-constrained values, which were comparable to the values of the published average C:N ratios values across the world in the 1990s (C:P = 13:1~60:1, N:P = 0.9:1~5:1). Generally, the soil C:N and C:P ratios experienced consistent increases in the Taihu Lake basin from the 1980s to 2000 due partly to the continuous straw return practices. The regions with fast economic growth in the upstream of the Taihu Lake basin generally experienced increases in soil C:N, C:P, and N:P due to the reduction in agricultural land. The changes in LUTs and their corresponding management practices shaped the spatial and temporal characteristics of soil C:N:P in the Taihu Lake basin. The results suggested that the imbalanced C:N:P stoichiometry characteristics of agricultural soil may lead to decoupled C, N, and P biogeochemical cycles, thereby having an adverse impact on soil carbon sequestration.

Supplementary Materials: The following supporting information can be downloaded at: https://www.mdpi.com/article/10.3390/agriculture13020484/s1, Table S1: Agricultural land area changes in different land use type during the periods of 1980–2000 and 2000–2016 in Tai Lake basin; Figure S1. Temporal and spatial distribution of top soil C, N and P in 1980s and 2000s, respectively, in Taihu Lake basin; Figure S2. Changes of top soil C:N, C:P and N:P ratio between 1980s and 2000s in Taihu Lake basin. The expression unit of the changes is "%".

Author Contributions: Conceptualization, C.L. and Z.Y.; methodology, C.L. and Z.Y.; formal analysis, C.L. and Z.Y.; investigation, C.L. and Z.Y.; resources, Z.Y. and Y.D.; data curation, Y.D., C.L. and Z.Y.; writing—original draft preparation, C.L. and J.L.; writing—review and editing, P.L., J.L. and C.L.; supervision, Y.G. and P.L.; visualization, J.L.; project administration, W.S., Y.G. and J.L.; funding acquisition, W.S., Y.G. and J.L. All authors have read and agreed to the published version of the manuscript.

Funding: This research was jointly supported by the Jiangsu R & D Special Fund for Carbon Peaking and Carbon Neutrality (grant number: BK20220014), the National Natural Science Foundation of China (41571458, 41907026, 41871209), the China Postdoctoral Science Foundation (2018M632253), and the Open Foundation of Hebei Key Laboratory of Wetland Ecology and Conservation (hklk202005; hklz201904).

Institutional Review Board Statement: Not applicable.

Data Availability Statement: All data are fully available without restriction.

Acknowledgments: We extend a special thanks to Hongjie Di (College of Agriculture and Life Sciences, Lincoln University) for English editing and Yuhan Niu (Co-Innovation Center for Sustainable Forestry in Southern China, Nanjing Forestry University) for the literature download from the China National Knowledge Infrastructure (CNKI) database and ISI-Web of Science.

Conflicts of Interest: The authors declare no conflict of interest.

References

1. Schimel, W.J.P. Interactions between Carbon and Nitrogen Mineralization and Soil Organic Matter Chemistry in Arctic Tundra Soils. *Ecosystems* **2003**, *6*, 129–143.
2. Graaff, M.A.D.; Kessel, C.V.; Six, J. The impact of long-term elevated CO_2 on C and N retention in stable SOM pools. *Plant Soil* **2008**, *303*, 311–321. [CrossRef]
3. Yu, Z.; Chen, L.; Pan, S.; Li, Y.; Kuzyakov, Y.; Xu, J.; Brookes, P.C.; Luo, Y. Feedstock determines biochar-induced soil priming effects by stimulating the activity of specific microorganisms. *Eur. J. Soil Sci.* **2018**, *69*, 521–534. [CrossRef]
4. Ciais, P.; Sabine, C.; Bala, G.; Bopp, L.; Brovkin, V.; Canadell, J.; Chhabra, A.; DeFries, R.; Galloway, J.; Heimann, M.; et al. Carbon and other biogeochemical cycles. In *Climate Change 2013: The Physical Science Basis*; Contribution of Working Group I to the Fifth Assessment Report of the Intergovernmental Panel on Climate Change; Cambridge University Press: Cambridge, UK, 2014; pp. 465–570.
5. Lorenz, K.; Lal, R. Soil organic carbon sequestration in agroforestry systems. A review. *Agric. Sustain. Dev.* **2014**, *34*, 443–445. [CrossRef]
6. Luo, Z.K.; Wang, E.L.; Baldock, J.; Xin, H.T. Potential soil organic carbon stock and its uncertainty under various cropping systems in Australian cropland. *Soil Res.* **2014**, *52*, 463–475. [CrossRef]
7. Li, Y.H.; Sun, H.F.; Li, H.R.; Yang, L.S.; Ye, B.X.; Wang, W.Y. Dynamic changes of rhizosphere properties and antioxidant enzyme responses of wheat plants (*Triticum aestivum* L.) grown in mercury-contaminated soils. *Chemosphere* **2013**, *93*, 972–977. [CrossRef]
8. Guo, P.D.; Sun, Y.S.; Su, H.T.; Wang, M.X.; Zhang, Y.X. Spatial and temporal trends in total organic carbon (TOC), black carbon (BC), and total nitrogen (TN) and their relationships under different planting patterns in a restored coastal mangrove wetland: Case study in Fujian, China. *Chem. Spec. Bioavailab.* **2018**, *30*, 1–10. [CrossRef]
9. Yuan, H.; Liu, S.; Razavi, B.S.; Zhran, M.; Wang, J.; Zhu, Z.; Wu, J.S.; Ge, T. Differentiated response of plant and microbial C: N: P stoichiometries to phosphorus application in phosphorus-limited paddy soil. *Eur. J. Soil Biol.* **2019**, *95*, 103–122. [CrossRef]
10. Cleveland, C.C.; Liptzin, D. C:N:P Stoichiometry in Soil: Is There a "Redfield Ratio" for the Microbial Biomass? *Biochemistry* **2007**, *85*, 235–252. [CrossRef]
11. Zhao, F.; Sun, J.; Ren, C.; Kang, D.; Deng, J.; Han, X.; Yang, G.; Feng, Y.; Ren, G. Land use change influences soil C, N, and P stoichiometry under "Grain-to-Green Program" in China. *Sci. Rep.* **2015**, *5*, 10195.
12. Zhu, Z.; Ge, T.; Luo, Y.; Liu, S.; Xu, X.; Tong, C.; Shibistova, O.; Guggenberger, G.; Wu, J. Microbial stoichiometric flexibility regulates rice straw mineralization and its priming effect in paddy soil. *Soil Biol. Biochem.* **2018**, *121*, 67–76. [CrossRef]
13. Sun, J.; Gao, P.; Xu, H.; Li, C.; Niu, X. Decomposition dynamics and ecological stoichiometry of Quercus acutissima and Pinus densiflora litter in the Grain to Green Program Area of northern China. *J. For. Res.* **2020**, *31*, 1613–1623. [CrossRef]
14. Duan, Y.; Chen, L.; Li, Y.M.; Wang, Q.Y.; Zhang, C.Z.; Ma, D.H.; Li, J.Y.; Zhang, J.B. N, P and straw return influence the accrual of organic carbon fractions and microbial traits in a Mollisol. *Geoderma* **2021**, *403*, 115–373. [CrossRef]
15. Maynard, D.G.; Paré, D.; Thiffault, E.; Lafleur, B. How do natural disturbances and human activities affect soils and tree nutrition and growth in the Canadian boreal forest? *Environ. Rev.* **2014**, *22*, 161–178. [CrossRef]
16. Aminiyan, M.M.; Aminiyan, F.M.; Mousavi, R.; Heydariyan, A. Heavy metal pollution affected by human activities and different land-use in urban topsoil: A case study in Rafsanjan city, Kerman province, Iran. *Eur. J. Soil Sci.* **2016**, *5*, 97–104. [CrossRef]
17. Don, A.; Schumacher, J.; Freibauer, A. Impact of tropical land-use change on soil organic carbon stocks—A meta-analysis. *Glob. Chang. Biol.* **2011**, *17*, 1658–1670. [CrossRef]
18. Gasparri, N.I.; Grau, H.R.; Manghi, E. Carbon Pools and Emissions from Deforestation in Extra-Tropical Forests of Northern Argentina Between 1900 and 2005. *Ecosystems* **2008**, *11*, 1247–1261. [CrossRef]
19. Murty, D.; Kirschbaum, M.U.F.; Mcmurtrie, R.E.; Mcgilvray, H. Dose conversion of forest to agricultural land change soil carbon and nitrogen? A review of the literature. *Glob. Chang. Biol.* **2002**, *8*, 105–123. [CrossRef]
20. Volante, J.N.; Alcarazsegura, D.; Mosciaro, M.J.; Viglizzod, E.F.; Paruelob, J.M. Ecosystem functional changes associated with land clearing in NW Argentina. *Agric. Ecosys. Environ.* **2012**, *154*, 12–22. [CrossRef]
21. Xia, X.; Zhang, P.; He, L.; Gao, X.; Li, W.; Zhou, Y.; Li, Z.X.; Li, H.; Yang, L. Effects of tillage managements and maize straw returning on soil microbiome using 16S rDNA sequencing. *J. Integr. Plant Biol.* **2019**, *61*, 765–777. [CrossRef]
22. Xu, Y.; Ge, Y.; Lou, Y.; Meng, J.; Shi, L.; Xia, F. Assembly strategies of the wheat root-associated microbiome in soils contaminated with phenanthrene and copper. *J. Hazard. Mater.* **2021**, *412*, 125340. [CrossRef] [PubMed]
23. Xu, X.; Liu, Y.; Singh, B.P.; Yang, Q.; Zhang, Q.; Wang, H.; Xia, Z.; Di, H.; Singh, B.K.; Xu, J.; et al. NosZ clade II rather than clade I determine in situ N_2O emissions with different fertilizer types under simulated climate change and its legacy. *Soil Biol. Biochem.* **2020**, *150*, 107974. [CrossRef]
24. Falkengren-Grerup, U.; Brink, D.J.T.; Brunet, J. Land use effects on soil N, P, C and pH persist over 40–80 years of forest growth on agricultural soils. *For. Ecol. Manag.* **2006**, *225*, 74–81. [CrossRef]
25. Bai, Z.; Wu, X.; Lin, J.J.; Xie, H.T.; Yuan, H.S.; Liang, C. Litter-, soil- and C:N-stoichiometry-associated shifts in fungal communities along a subtropical forest succession. *Catena* **2019**, *178*, 350–358. [CrossRef]
26. Sardans, J.; Peñuelas, J. Tree growth changes with climate and forest type are associated with relative allocation of nutrients, especially phosphorus, to leaves and wood. *Glob. Ecol. Biogeogr.* **2013**, *22*, 494–507. [CrossRef]
27. Mulder, C.; Elser, J.J. Soil acidity, ecological stoichiometry and allometric scaling in grassland food webs. *Glob. Chang. Biol.* **2009**, *15*, 2730–2738. [CrossRef]

28. Peñuelas, J.; Sardans, J.; Alcaniz, J.M.; Poch, J.M. Increased eutrophication and nutrient imbalances in the agricultural soil of NE Catalonia, Spain. *J. Environ. Biol.* **2009**, *30*, 841–846.
29. Peñuelas, J.; Poulter, B.; Sardans, J.; Ciais, P.; van der Velde, M.; Bopp, L.; Boucher, O.; Godderis, Y.; Llusià, J.; Nardin, E.; et al. Human-induced nitrogen–phosphorus imbalances alter natural and managed ecosystems across the globe. *Nat. Commun.* **2013**, *4*, 1–10. [CrossRef]
30. Bell, C.; Carrillo, Y.; Boot, C.M.; Rocca, J.D.; Pendal, E.; Wallenstein, M.D. Rhizosphere stoichiometry: Are C:N:P ratios of plants, soils, and enzymes conserved at the plant species-level? *New Phytol.* **2014**, *201*, 505–517. [CrossRef]
31. Heyburn, J.; Mckenzie, P.; Crawley, M.J. Effects of grassland management on plant C:N:P stoichiometry: Implications for soil element cycling and storage. *Ecosphere* **2017**, *8*, 1–14. [CrossRef]
32. Shao, Y.L.; Xu, Y.P.; Ma, S.S. The Study of Land Use Change on the Landscape Impact of Urban Water in Taihu Lake Basin—A Case Study in Urban Suzhou. *Appl. Mechan. Mater.* **2012**, *209–211*, 325–330. [CrossRef]
33. Zhao, H.X.; You, B.S.; Duan, X.J.; Becky, S. Industrial and agricultural effects on water environment and its optimization in heavily polluted area in Taihu Lake Basin, China. *Chin. Geogr. Sci.* **2013**, *23*, 203–215. [CrossRef]
34. Pan, G.X.; Zhao, Q.G. Study on evolution for carbon stock in agricultural soil of China: Facing the challenge of global change and food security. *Adv. Earth Sci.* **2005**, *20*, 284–393. (In Chinese)
35. Zhao, X.; Xie, Y.X.; Xiong, Z.Q.; Yan, X.Y. Nitrogen fate and environmental consequence in paddy soil under rice–wheat rotation in the Taihu lake region, China. *Plant Soil* **2009**, *319*, 225–234. [CrossRef]
36. Yang, Y.G.; Shen, Y.; Shao, S.G.; Li, X.L. Spatial distribution in forest soil nutrients and its relationship with ecological factors on the up stream of Taihu Lake Basin. *Adv. Mater. Res.* **2013**, *652–654*, 1660–1663. [CrossRef]
37. Chu, J.; Xue, J.H.; Jin, M.J.; Yong, B.; Shi, H.; Xu, Y.Q. Effects of poplar-wheat intercropping system on soil nitrogen loss in Taihu Basin. *Trans. Chin Soc. Agric. Eng.* **2015**, *31*, 167–177. (In Chinese)
38. Sommers, L.E.; Nelson, D.W. Determination of total phosphorus in soils: A rapid perchloric acid digestion procedure 1. *Soil Sci. Soc. Am. J.* **1972**, *36*, 902–904. [CrossRef]
39. Ning, Q.; Chen, L.; Jia, Z.J.; Zhang, C.Z.; Ma, D.H.; Li, F.; Zhang, J.B.; Li, D.M.; Han, X.R.; Cai, Z.J.; et al. Multiple long-term observations reveal a strategy for soil pH-dependent fertilization and fungal communities in support of agricultural production. *Agric. Ecosyst. Environ.* **2020**, *293*, 106–837. [CrossRef]
40. Williams, A.; Borjesson, G.; Hedlund, K. The effects of 55 years of different inorganic fertiliser regimes on soil properties and microbial community composition. *Soil Biol. Biochem.* **2013**, *67*, 41–46. [CrossRef]
41. Tian, H.; Chen, G.; Zhang, C.; Melillo, J.; Hall, C. Pattern and variation of C:N:P ratios in China's soils: A synthesis of observational data. *Biogeochemistry* **2010**, *98*, 139–151. [CrossRef]
42. Bui, E.N.; Henderson, B.L. C:N:P stoichiometry in Australian soils with respect to vegetation and environmental factors. *Plant Soil* **2013**, *373*, 553568. [CrossRef]
43. Wang, W.; Xie, X.; Chen, A.; Yin, C.; Chen, W. Effects of long-term fertilization on soil carbon, nitrogen, phosphorus and rice yield. *J. Plant Nutr.* **2013**, *36*, 551–561. [CrossRef]
44. Schleuss, P.M.; Widdig, M.; Heintz-Buschart, A.; Guhr, A.; Martin, S.; Kirkman, K.; Spohn, M. Stoichiometric controls of soil carbon and nitrogen cycling after long-term nitrogen and phosphorus addition in a mesic grassland in South Africa. *Soil Biol. Biochem.* **2019**, *135*, 294–303. [CrossRef]
45. Jiang, X.S.; Liu, C.L.; Sui, B.; Dong, C.X.; Guo, S.W. Problems and Proposals of the Current Fertilization Situation in the rice–wheat Rotation System in Tai Lake Basin. *Chin. Agric. Sci. Bull.* **2012**, *28*, 15–18. (In Chinese)
46. Jin, Z.Q.; Shah, T.; Zhang, L.; Liu, H.Y.; Peng, S.B.; Nie, L.X. Effect of straw returning on soil organic carbon in rice–wheat rotation system: A review. *Food Energy Secur.* **2020**, *9*, e200. [CrossRef]
47. Li, Y.E.; Shi, S.; Waqas, M.A.; Zhou, X.; Li, J.; Wan, Y.; Wilkes, A. Long-term ($\geq$20 years) application of fertilizers and straw return enhances soil carbon storage: A meta-analysis. *Mitig. Adapt. Strateg. Glob. Chang.* **2018**, *23*, 603–619. [CrossRef]
48. Lu, F.E.I.; Wang, X.; Han, B.; Ouyang, Z.; Duan, X.; Zheng, H.U.A.; Miao, H. Soil carbon sequestrations by nitrogen fertilizer application, straw return and no-tillage in China's cropland. *Glob. Chang. Biol.* **2009**, *15*, 281–305. [CrossRef]
49. Tang, Q.; Cotton, A.; Wei, Z.; Xia, Y.; Daniell, T.; Yan, X. How does partial substitution of chemical fertiliser with organic forms increase sustainability of agricultural production? *Sci. Total Environ.* **2022**, *803*, 149933. [CrossRef]
50. Raiesi, F.; Kabiri, V. Carbon and nitrogen mineralization kinetics as affected by tillage systems in a calcareous loam soil. *Ecol. G* **2017**, *106*, 24–34. [CrossRef]
51. Kuzyakov, Y.; Friedel, J.K.; Stahr, K. Review of mechanisms and quantification of priming effects. *Soil Biol. Biochem.* **2000**, *32*, 1485–1498. [CrossRef]
52. Delgado-Baquerizo, M.; Maestre, F.T.; Gallardo, A. Biological soil crusts increase the resistance of soil nitrogen dynamics to changes in temperatures in a semi-arid ecosystem. *Plant Soil* **2013**, *366*, 35–47. [CrossRef]

Article

Altered Organic Matter Chemical Functional Groups and Bacterial Community Composition Promote Crop Yield under Integrated Soil–Crop Management System

Qi Li [1,2], Amit Kumar [3], Zhenwei Song [4], Qiang Gao [1], Yakov Kuzyakov [5], Jing Tian [2,*] and Fusuo Zhang [2]

[1] College of Resources and Environment, Jilin Agricultural University, Changchun 130118, China
[2] College of Resources and Environmental Sciences, Key Laboratory of Plant-Soil Interactions, Ministry of Education, National Academy of Agriculture Green Development, China Agricultural University, Beijing 100193, China
[3] Biology Department, College of Science, United Arab Emirates University, Al Ain P.O. Box 15551, United Arab Emirates
[4] Institute of Crop Sciences, Chinese Academy of Agricultural Sciences/Key Laboratory of Crop Physiology and Ecology, Ministry of Agriculture and Rural Affairs of China, Beijing 100081, China
[5] Department of Soil Science of Temperate Ecosystems, University of Göttingen, 37077 Göttingen, Germany
* Correspondence: tianj@igsnrr.ac.cn; Tel.: +86-010-62734554

Citation: Li, Q.; Kumar, A.; Song, Z.; Gao, Q.; Kuzyakov, Y.; Tian, J.; Zhang, F. Altered Organic Matter Chemical Functional Groups and Bacterial Community Composition Promote Crop Yield under Integrated Soil–Crop Management System. *Agriculture* **2023**, *13*, 134. https://doi.org/10.3390/agriculture13010134

Academic Editor: Ryusuke Hatano

Received: 18 November 2022
Revised: 30 December 2022
Accepted: 31 December 2022
Published: 4 January 2023

Abstract: Sustainable agricultural production is essential to ensure an adequate food supply, and optimal farm management is critical to improve soil quality and the sustainability of agroecosystems. Integrated soil–crop management based on crop models and nutrient management designs has proven useful in increasing yields. However, studies on its effects on the chemical composition of soil organic carbon (SOC) and microbial community composition, as well as their linkage with crop yield, are lacking. Here, we investigated the changes in SOC content, its chemical functional groups, and bacterial communities, as well as their association with crop yield under different farmland management based on four farmland management field trials over 12 years (i.e., FP: farmer practice; IP: improved farmer practice; HY: high-yield system; and ISSM: integrated soil–crop system management). The crop yield increased by 4.1–9.4% and SOC content increased by 15–87% in ISSM compared to other farmland management systems. The increased proportion of Methoxy C and O-alkyl C functional groups with a low ratio of Alkyl C/O-alkyl C, but high Aliphatic C/Aromatic C in ISSM hints toward slow SOC decomposition and high soil C quality. The relative abundances of r-strategists (e.g., Firmicutes, Myxobacteria, and Bacteroidetes) was highest under the ISSM. Co-occurrence network analysis revealed highly complex bacterial communities under ISSM, with greater positive links with labile SOC functional groups. The soil fertility index was the main factor fueling crop yields, as it increased with the relative abundance of r-strategists and SOC content. Our results indicated that crop yield advantages in ISSM were linked to the high C quality and shifts in bacterial composition toward r-strategists by mediating nutrient cycling and soil fertility, thereby contributing to sustainability in cropping systems.

Keywords: SOC functional group; bacterial community structure; crop yield; soil fertility; integrated soil–crop management system

1. Introduction

Soil organic matter (SOM) supports multiple ecosystem functions and is closely linked to consistently high crop yields and sustainable agriculture [1,2]. Therefore, maintaining and enhancing SOM content in agricultural fields is important, especially within the context of food security and climate change mitigation [3,4]. Although agricultural land covers 38% of the Earth's land surface, large areas of agricultural land suffer from medium to strong degradation [5]. Therefore, improving soil organic carbon (SOC) is an efficient way to increase C sinks, reduce greenhouse gas (CO_2) emissions, and improve crop productivity in carbon-depleted agricultural soils [6,7].

Inappropriate management has depleted 25–75% of SOC in global cropland [8], resulting in a growing interest in optimizing agricultural management. Current optimization techniques include judicious use of fertilizer, adding manure, or returning straw to the field [9–13]. Straw returns maximize the use of natural resources, improve soil structure, and increase the SOM content, providing an ideal environment for crop plantations [14,15]. Organic fertilizers are rich in organic matter and beneficial microorganisms [16]. Many field trials have shown that long-term organic fertilization can increase the conversion of organic and inorganic materials in the soil, which is conducive to improving soil fertility and promoting crop growth and development. [12,17–19].

SOM is chemically diverse and consists of pools with various availability and turnover rates [20,21]. The chemical composition of SOM, which defines its decomposability and stability, is receiving attention in the context of research on soil fertility and quality [22,23]. Intensive management has a direct and indirect impact on the quality and quantity of organic C entering agricultural soils. For instance, organic fertilization generally increases Alkyl C and aromatic C, while decreasing O-alkyl C in soil [23,24]. On the contrary, combined mineral NPK fertilizers (combined with organic manure) increase the O-alkyl group levels but decrease the levels of Alkyl and aromatic groups and the ratio of Alkyl-C/O-alkyl-C (A/O-A) groups [25,26]. The inconsistent compositions of SOM chemical functional groups could be attributable to site-specific soil conditions, input types, climate, and the complexity of the microbial decomposition processes [27–30].

SOC is considered an overarching edaphic factor that shapes bacterial diversity [12,31,32]. This is because heterotrophic soil microorganisms rely on increased SOC to obtain their nutrients [33–35]. After long-term organic matter input, greater SOC is observed, and it is accompanied by greater microbial activity [36,37]. Specifically, when SOC increases, there tends to be a shift in microbial community composition; the community moves toward having more *r*-strategists compared to unfertilized plots [38]. Beyond determining life history strategies, SOC also determines which bacterial phyla are present in the soil [31,32]; this may result from the ability of the microbial groups to utilize SOC pools [39]. Long-term organic amendments (e.g., sewage sludge and chicken manure) increase the dissolved organic carbon content and support an indicative shift of bacterial community to r-strategist taxa such as Proteobacteria and Actinobacteria [40]. This is because the r-strategists have high nutrient requirements and can maximize their reproductive outputs when resources are abundant [31]. The application of organic fertilizers mitigates the negative effects of mineral fertilization on microbial diversity, and the shift of microbial communities toward *r*-strategists is owing to the efficient input of C that characterizes organic fertilizers [36–38]. All these studies suggest that SOC quality is vital for adjusting microbial community composition and life history.

To increase crop production while minimizing impact on the environment, the integrated soil crop system management (ISSM) strategy was recently introduced. The strategy is based on a Hybrid-Maize simulation model and a seasonal root zone nitrogen management strategy (IRNM) to determine the most appropriate combination of planting date, crop density, and fertilizer application rate at the trial site [41]. ISSM has been shown to increase yields (maize, rice, and wheat) by an average of 10.8–11.5% from 2005–2015 [41–43].

Northeast China is considered the "first granary", with grain production accounting for a quarter of the country's total grain production [44]. SOC levels in Northeast China have declined by 22% over the past three decades [2,45], and the decrease is mainly due to intensive cropping and inefficient fertilization [46,47]. Therefore, it is crucial to identify appropriate soil management practices that are conducive to soil C sequestration and sustainable development of agricultural ecosystems in this region. In this study, we aimed to (i) compare the changes in total SOC content, chemical functional groups, and bacterial community composition across various farmland management systems; and (ii) explore changes in SOC quantity and quality in relation to bacterial communities and their impact on crop yield. We hypothesize that (1) compared to farmer practice (FP), ISSM will increase SOC, especially the labile functional pools, because of the direct input of nutrients associated with this strategy [48]; (2) changes in SOC quantity and quality in the ISSM system are important factors shaping microbial community composition, with increasing r-strategist bacteria ascribed to preferentially exploit labile organic compounds [49,50]; (3) changes in SOC quantity and quality impact crop yield via regulating bacterial community composition and enhancing soil fertility [18].

2. Materials and Methods

2.1. Study Site

The field experiment was initiated in April 2009 using a maize monoculture, in Zhuling Gong County (43°12' N, 124°66' E, 206 m elevation), Ji Lin Province, Northeast China. The site has a temperate continental monsoon climate, an annual precipitation of 595 mm, and an annual mean temperature of 6.3 °C. The soil in this region is classified as Phaeozem (equivalent to Hapudoll as per USDA Soil Taxonomy). At the start of the experiment, the topsoil (0–20 cm) had the following basic characteristics: pH, 6.7; SOC, 16 g kg^{-1}; total N (TN), 1.6 g kg^{-1}; available P (AP), 7 mg kg^{-1}; and available K (AK), 151 mg kg^{-1}. Four farmland management systems were established: (1) farmer practice (FP)—set up based on a survey of various production management systems that local farmers are accustomed to applying; (2) improved farmer practice (IP)—a model based on farmer practice by optimizing the combination of existing techniques (e.g., planting density and fertilizer application) to increase crop yields by 15–20%; (3) high-yield system (HY)—a management model intended to increase crop yield by 30%, regardless of the associated environmental costs; and (4) integrated soil–crop system management (ISSM)—a sustainable integrated program to increase crop yield and resource use efficiency by optimizing planting density, fertilizer application, and tillage systems with the aim of achieving an ultrahigh yield while lowering resource and environmental costs.

The field trial design was a randomized complete block design with four farmland management treatments and four replications, for a total of 16 plots. The size of each plot was 6 m × 23 m. For each farmland management system, details of the fertilization rate, planting density, and straw management are shown in Table 1. Maize (ZhongDan99 cultivar) was sown in late April and harvested in early October. The straw was returned to the field based on the amount of maize root and shoot (leaf and stem) residues after above-ground harvesting in each plot, representing 30% of the total dry matter weight.

Table 1. The detailed field management information for four farmland management systems.

Management Systems	N (kg·ha^{-1})	P$_2$O$_5$ (kg·ha^{-1})	K$_2$O (kg·ha^{-1})	Organic Fertilizers (kg·ha^{-1})	Straw Management	Planting Density (Plants·ha^{-1})
Farmer practice (FP)	300	120	120	-	30%	50,000
Improved farmer practice (IP)	195	90	90	-	30%	60,000
High-yield system (HY)	300	120	120	-	30%	70,000
Integrated soil–crop system management (ISSM)	195	90	90	10,000	30%	70,000

FP: farmer practice; IP: improved farmer practice; HY: high-yield system; ISSM: integrated soil–crop system management. Straw management: amount of straw returned to the field.

2.2. Soil Sampling and Analysis of Physicochemical Properties

Five topsoil cores (0–20 cm depth, 5 cm diameter) were collected at random from each plot in October 2020 and pooled together to form one composite sample per plot, for a total of 16 soil samples. After that, the composite samples were passed through a 2 mm sieve to remove any roots or stones. Each soil sample was stored in three parts: one part was air-dried and stored at room temperature for the determination of various chemical properties, e.g., SOC, TN, pH, AP, and AK; the second sample was stored in a refrigerator at

−20 °C for the determination of ammonium (NH_4^+) and nitrate (NO_3^-) content; the third part was stored at −80 °C and then subjected to DNA extraction and microbial community analysis. Soil samples were collected from 0–20 cm depths using cutting rings of 100 cm^3 volume to calculate soil bulk density (BD). All maize plants in each plot were collected at the ripening stage, and the crop yield was analyzed as the dry weight of grains.

The SOC and TN contents were determined through combustion using a Vario EL III Elemental Analyzer (Elementar). Soil available phosphorus (AP) and available potassium (AK) were measured according to Hanway and Heidal (1952) and Olsen (1954) [51,52], respectively. Soil pH was measured using a pH electrode (FE20-FiveEasy pH, Mettler Toledo) with a soil-to-deionized-water ratio of 1:2.5 (w/w). Soil mineral N, including NO_3^- and NH_4^+, was extracted with 0.01 mol L^{-1} KCl solution and analyzed using an auto-analyzer (TRAACS-2000, Bran+Luebbe).

The SOC stock was calculated using the following equation [53]:

$$SOC\ stocks\ \left(\text{Mg C ha}^{-1}\right) = SOC\left(\text{g kg}^{-1}\right) \times BD\left(\text{g m}^{-3}\right) \times \frac{H(\text{cm})}{100} \tag{1}$$

where BD and H represent bulk density and soil depth (0–20 cm), respectively.

The soil fertility index was determined using the averaging approach as follows: to prevent the differences in data scale, TN, AP, AK, NO_3^-, and NH_4^+ estimated in this investigation were normalized from 0 to 1 using the "max–min" approach, and the soil fertility index was then calculated as follows [54]:

$$Fertility\ index = (-x_{i,min})/(x_{i,max} - x_{i,min}) \tag{2}$$

where x_i is the measured soil properties, $x_{i,min}$ is the minimum of soil properties I, $x_{i,max}$ is the maximum of soil properties i.

2.3. Solid-State ^{13}C NMR Spectroscopy

The chemical composition of SOM was investigated by determining the relative abundance of functional groups using solid-state ^{13}C cross-polarization and magic-angle-spin (CPMAS) NMR. Soil samples were treated with a 10% hydrofluoric acid (HF) solution to concentrate the organic matter and remove paramagnetic minerals [55]. The ^{13}C-CPMAS NMR spectra were acquired using an AVANCE III 400 WB spectrometer (Bruker, Billerica, MA, USA) at 100 MHz for ^{13}C and 400 MHz for ^{1}H with a spinning rate of 5 kHz, an acquisition of 20 ms, a recycle time of 1 s, and a contact time of 1 ms.

Bruker TopSpin (v4.1.1) was used to compare the ^{13}C-CPMAS NMR spectra of different samples; peak areas were calculated and integrated to estimate their relative proportions. The methods outlined by Bonanoni et al. were used to select spectral regions and identify C functional groups (chemical structures) [56]: 0–50 ppm, Alkyl C; 50–60 ppm, Methoxyl C; 60–95 ppm, O-alkyl C; 95–110 ppm, Di-O-alkyl C; 110–145 ppm, Aryl C; 145–160 ppm, Phenolic C; and 160–190 ppm, Carboxyl C. In this study, the Methoxyl C, O-alkyl C, Di-O-alkyl C, and Carboxyl C groups were classified as labile C groups, whereas Alkyl C, Aryl C, and Phenolic C groups were classified as recalcitrant C groups [57,58]. Different indices of SOM stability were calculated as follows [59,60]:

The *A/O-A* (*Alkyl C/O-alkyl C*) represents the degree of decomposition and was calculated as follows:

$$A/O\text{-}A\ ratio = \frac{Alkyl\ C}{Methoxyl\ C + O\text{-}alkyl\ C + Di\text{-}O\text{-}alkyl\ C} \tag{3}$$

The Aliphatic C/Aromatic C (*Alip/Arom*) indicates the degree of aliphaticity, and was calculated using the following equation:

$$Alip/Arom = \frac{Alkyl\ C + Methoxyl\ C + O\text{-}alkyl\ C + Di\text{-}O\text{-}alkyl\ C}{Aryl\ C + Phenolic\ C} \tag{4}$$

The *HB/HI* (Hydrophobic C/Hydrophilic C) indicates the hydrophobicity degree, and was calculated using the following equation:

$$HB/HI = \frac{Alkyl\ C + Aryl\ C + Phenolic\ C}{Methoxyl\ C + O\text{-}Alkyl\ C + Di\text{-}O\text{-}alkyl\ C + Carboxyl\ C} \tag{5}$$

The *Aromaticity* response to soil SOC stability was calculated as follows:

$$Aromaticity = \frac{Aryl\ C + Phenolic\ C}{Alkyl\ C + Methoxyl\ C + O\text{-}alkyl\ C + Di\text{-}O\text{-}alkyl\ C} \tag{6}$$

2.4. DNA Extraction and Amplicon Sequencing

DNA was extracted from fresh soil (0.25 g) according to the manufacturer's instructions using a PowerSoil kit (MoBio Laboratories, Carlsbad, CA, USA). The DNA extract was tested on a 1% agarose gel, and the concentration and purity of the DNA were determined using a NanoDrop 2000 UV-vis spectrophotometer (ThermoScientific, Wilmington, USA). The hypervariable region V3-V4 of the bacterial 16S rRNA gene was amplified with primer pairs 338F (5′-ACTCCTACGGGAGGCAGCAG-3′) and 806R (5-GGACTACHVGGGTWTCTAAT-3′) using an ABI GeneAmp® 9700 PCR thermocycler (ABI, Carlsbad, CA, USA). The PCR amplification of the 16S rRNA gene was performed as follows: initial denaturation at 95 °C for 3 min, followed by 30 cycles of denaturing at 95 °C for 30 s, annealing at 55 °C for 30 s, and extension at 72 °C for 45 s, then single extension at 72 °C for 10 min, and a final hold at 10 °C. The PCR mixtures contained 4 μL of 5X TransStart FastPfu buffer, 2 μL of 2.5 mM dNTPs, 0.8 μL of forward primer (5 μM), 0.8 μL of reverse primer (5 μM), 0.4 μL of TransStart FastPfu DNA Polymerase, template DNA 10 ng, and finally ddH$_2$O up to 20 μL. The PCRs were performed in triplicate. The PCR product was extracted from 2% agarose gel and purified using the AxyPrep DNA Gel Extraction Kit (Axygen Biosciences, Union City, CA, USA) according to the manufacturer's instructions and quantified using Quantus™ Fluorometer (Promega, Madison, WI, USA).

Raw amplicon sequences were subjected to quality control using the following criteria: (1) the low-quality sequencing reads which had an average quality score of <20 or contained ambiguous nucleotides were filtered using Mohur version 1.31.1 using the UPARSE (version 7.1) software [61]; and (2) only overlapping sequences longer than 10 bp were assembled according to their overlapped sequence. The maximum mismatch ratio of the overlap region was 0.2. Reads that could not be assembled were discarded; sequences with ≥97% similarity were assigned to one operational taxonomic unit (OTU), and chimeric sequences were identified and removed. Taxonomy was assigned to each OTU by the RDP (Ribosomal Database Project) classifier [62]. The final sequencing products contained 43,750 sequences with an average length of 415 bp per sample for downstream analysis.

2.5. Statistical Analyses

SOC chemical functional groups, bacterial community composition, and other variables were analyzed by SPSS 22.0 (IBM Corp., Armonk, NY, USA) using one-way ANOVA with a randomized group design. Differences were considered significant at $p < 0.05$, and a post hoc least-significant-difference test was carried out to compare the differences among farmland management systems. The normal distribution of data was tested by using the "Shapiro. test" function of the stats package in R v.4.1.2 (R Core Team, 2021).

Changes in bacterial community composition were evaluated by nonmetric multidimensional scaling (NMDS), based on the Bray–Curtis distance calculation method. To further determine significant differences in bacterial community structure for any pair of samples, Adonis analysis was performed using NMDS and statistical analyses were performed using R (version 4.1.2) with the vegan package (R Core Team, 2021).

Linear discriminant analysis (LDA) effect size (LEfSe) was applied to determine if there were significant differences in bacterial taxa among the four farmland management systems [63]. We performed LDA in combination with Kruskal–Wallis (KW) test to identify species with significant differences in abundance between management systems setting log LDA scores > 2.0 (http://huttenhower.sph.harvard.edu/galaxy/, accessed on 16 November 2022). Circus (https://hiplot.com.cn/, accessed on 16 November 2022) was used to aid in the identification and analysis of similarities and differences resulting from phyla relative abundance comparisons [64], which emphasizes the statistical significance and biological relevance [63].

The co-occurrence networks of bacterial communities with SOC chemical functional groups and bacterial co-occurrence networks under each management system were compared by setting the same metrics (dissimilarity threshold for KLD matrix maxima and Spearman correlation threshold of 0.6) and running the networks using all OTU taxa for the four farmland management systems. For each edge and measure, the reciprocal and bootstrap distributions were generated with 100 iterations. The p-value of the measure was calculated as the area of the mean of the bootstrap distribution and the standard deviation generated by the mean of the Gaussian curve under the reciprocal distribution. The p-values were adjusted using the Benjamini–Hochberg procedure [65,66]. Only the most significant interactions with strong linear connections (r > 0.6) are shown in the network diagram. The nodes in the constructed network represent genus and SOC chemical functional groups, and the edges represent strong and significant correlations between the nodes. Network visualization was conducted using Gephi (version 0.9.2) and Cytoscape (version 3.8.2)(Ideker, 2011). The topological characteristics of the calculated network include positive and negative correlations, nodes, edges, network density, closeness centrality, and betweenness centrality (Tables S4 and S5).

To further elucidate the pathways through which all factors regulate crop yield, partial least-squares path modeling (PLS–PM) and Pearson correlations were performed. Model evaluation was performed based on the goodness of fit (GOF), with a GOF > 0.7 considered an acceptable value. The models were constructed using R (version 4.1.2) (R Core Team, 2021). Ordinary least-squares regression was performed to test the correlation between the factors.

3. Results

3.1. Soil Organic Matter, Soil Fertility, and Crop Yield

Compared with that in the FP, the SOC content increased by 15% in IP, 33% in HY, and 87% in ISSM ($p < 0.05$; Figure 1a). Correspondingly, the SOC stock increased by 11%, 31%, and 77% under IP, HY, and ISSM, respectively, compared to that in FP ($p < 0.05$; Figure 1b). Compared to the other three farmland management systems, ISSM increased TN, AP, AK, and NO_3^- contents ($p < 0.05$; Table 2). Consequently, ISSM increased the soil fertility ($p < 0.05$; Figure 1c) approximately four times than FP. The highest crop yield was obtained under ISSM, with a 9.4% increase than FP (Figure 1d).

Table 2. Effects of management systems on soil physicochemical properties in a 12-year field experiment.

Management Systems	pH	TN (g/kg)	SOC/TN	AP (mg kg^{-1})	AK (mg kg^{-1})	NO_3^- (mg N kg^{-1})	NH_4^+ (mg N kg^{-1})	BD (g m^3)
FP	5.04 ± 0.07 b	1.58 ± 0.10 b	10.6 ± 0.46 a	85.7 ± 3.08 b	223 ± 13.5 c	3.57 ± 0.39 b	6.32 ± 0.49 bc	1.23 ± 0.00 a
IP	5.27 ± 0.22 ab	1.69 ± 0.03 b	11.4 ± 0.31 a	84.6 ± 3.76 b	241 ± 16.1 bc	14.4 ± 0.59 a	8.94 ± 0.44 a	1.21 ± 0.03 a
HY	5.02 ± 0.12 b	1.96 ± 0.06 b	11.4 ± 0.31 a	105 ± 3.99 b	309 ± 8.35 b	11.1 ± 1.28 a	7.79 ± 0.38 ab	1.18 ± 0.01 a
ISSM	5.70 ± 0.10 a	2.87 ± 0.24 a	11.0 ± 0.59 a	175 ± 20.7 a	457 ± 18.8 a	14.8 ± 0.69 a	5.22 ± 0.55 c	1.16 ± 0.01 a

Error bars indicate standard errors of the mean (n = 4). Letters within the same row indicate significant differences among farmland management systems at $p < 0.05$. SOC: soil organic carbon; TN: total nitrogen; AP: available phosphorus; AK: available potassium; NH_4^+: ammonium N; NO_3^-: nitrate N; BD: soil bulk density. FP: farmer practice; IP: improved farmer practice; HY: high yield system; ISSM: Integrated soil–crop system management.

3.2. Soil Organic Matter Functional Groups Depending on Management Systems

Across all farmland management systems, the proportion of O-alkyl C (27–30%) was the largest, followed by Alkyl C (24–26%), Aryl C (16–21%), and Carboxyl C groups (7.9–9.2%). These four functional groups dominated SOM, accounting for more than 70% of the total functional group spectrum across all management systems (Table S1). Compared with FP, the ISSM increased the levels of Methoxyl C and O-alkyl C groups by 17% and 8.1%, respectively, and decreased Aryl C group levels by 21% ($p < 0.05$; Figure 2c,d). In general, the ISSM system increased the labile C groups ($p < 0.05$; Figure 2a). The A/O-A was similar in the FP and IP (0.55–0.60) and lower in ISSM (0.52) (Figure 2e). Among all management systems, the highest Alip/Arom was under ISSM (3.4) ($p < 0.05$; Figure 2e).

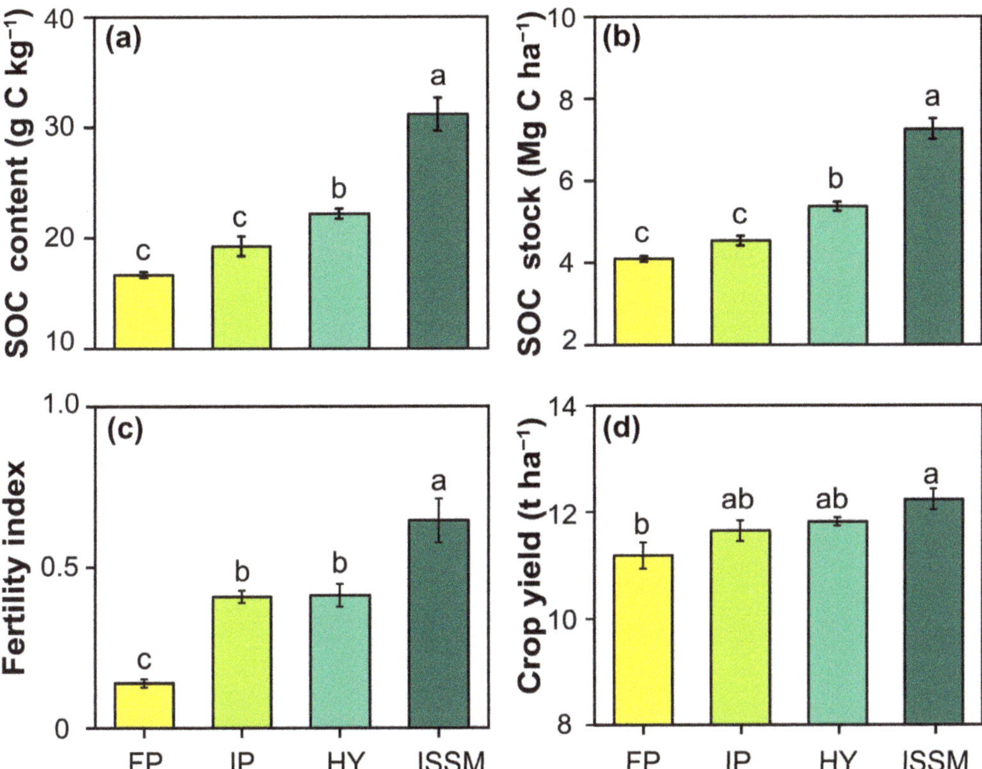

Figure 1. (**a,b**) Soil organic carbon content and stock (0–20 cm) under four farmland management systems, (**c,d**)soil fertility index, and crop yield after 12 years under four farmland management systems: FP: farmer practice; IP: improved farmer practice; HY: high-yield system; ISSM: integrated soil–crop system management. Letters represent significant differences ($p < 0.05$) among management systems. Error bars indicate standard errors of the mean (n = 4). The fertility index was calculated based on the standardized "max–min" approach of five soil properties.

3.3. Bacterial Community Structure Depending on Management Systems

The Shannon index of the bacterial community increased more in ISSM than in FP and HY ($p < 0.05$; Figure 3a). Management systems changed the soil bacteria community structure, as indicated by NMDS and Adonis tests ($p = 0.01$; Figure 3b). The pairwise comparison revealed that the soil bacterial community under ISSM was different from that under FP and HY (Table S3). A Mantel test revealed that the major factors shaping bacterial community structure were soil pH (r = 0.74, $p < 0.05$), AP (r = 0.67, $p < 0.05$), AK (r = 0.55, $p < 0.05$), SOC (r = 0.54, $p < 0.05$), TN (r = 0.40, $p < 0.05$), and O-alkyl C (r = 0.34, $p < 0.05$) (Table S4).

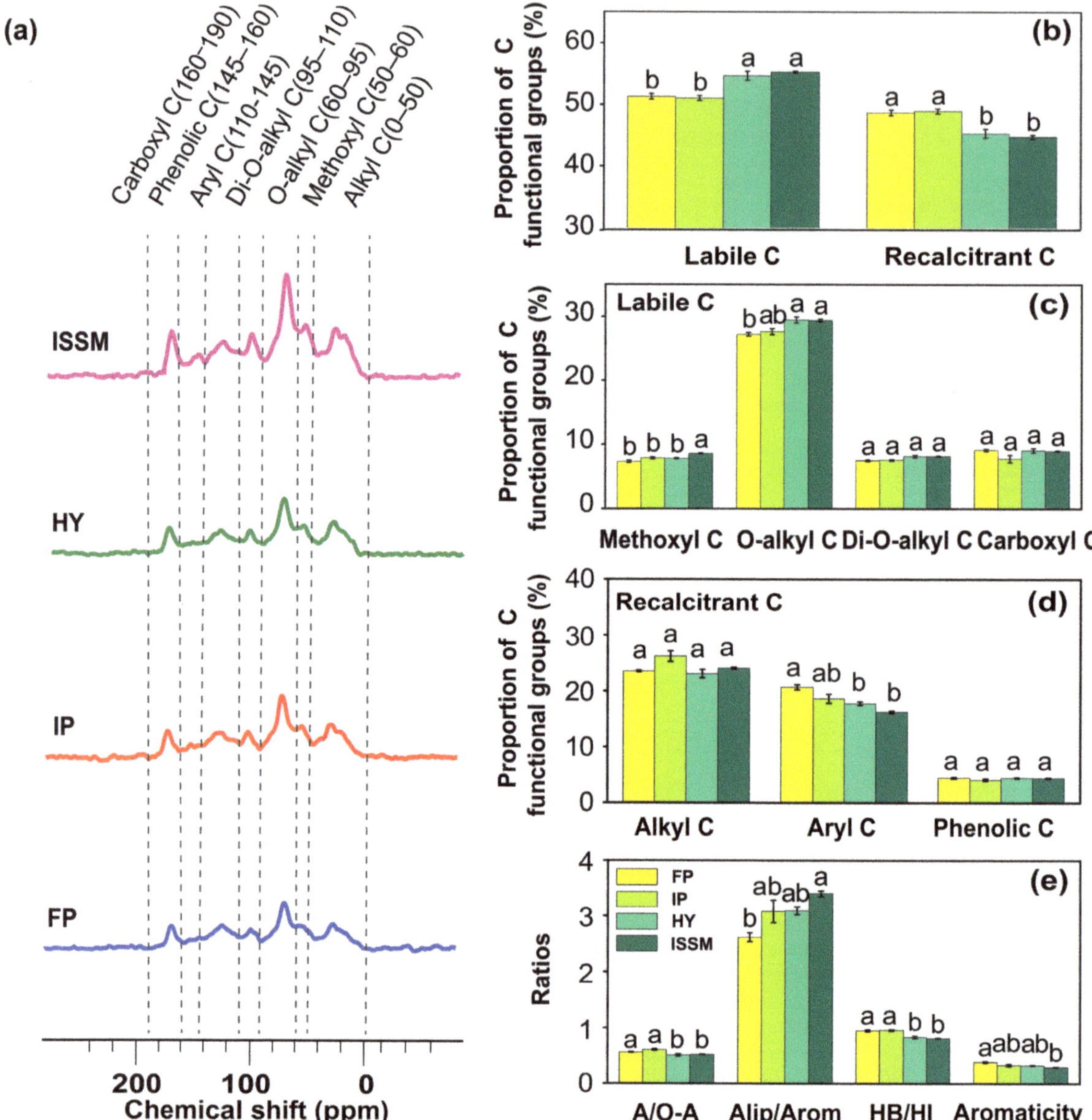

Figure 2. ^{13}C CPMAS NMR spectra of functional groups of SOC (**a**), and their proportions (**b–d**) and ratios (**e**) under four farmland management systems: FP: farmer practice; IP: improved farmer practice; HY: high-yield system; ISSM: integrated soil–crop system management. Letters represent significant differences ($p < 0.05$) among farmer practices. Error bars indicate standard errors of the mean ($n = 4$). A/O-A, Alkyl C/O-alkyl C; Alip/Arom, Aliphatic C/Aromatic C; HB/HI, Hydrophobic C/Hydrophilic C.

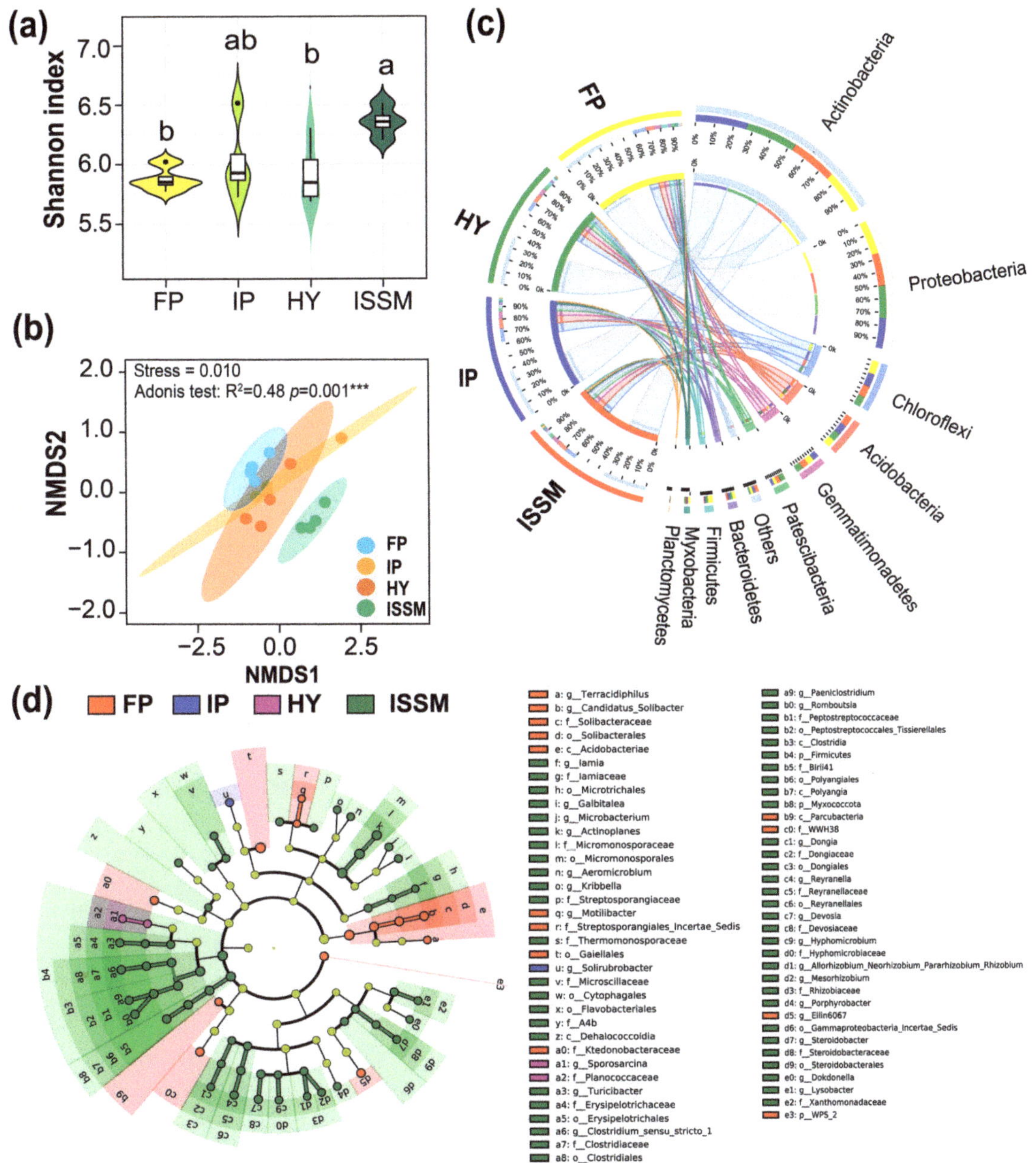

Figure 3. Bacterial community profiles in soil under four farmland management systems: FP: farmer practice; IP: improved farmer practice; HY: high-yield system; ISSM: integrated soil–crop system management. (**a**) Bacterial alpha-diversity (Shannon), (**b**) nonmetric multidimensional scaling (NMDS) analysis of bacterial community structure based on Bray–Curtis distance, (**c**) relative abundances of bacterial taxa at the phylum level, (**d**) the linear discriminant analysis effect size (LEfSe) analysis showed the significantly different taxa of bacterial communities. Letters represent significant differences ($p < 0.05$) among four farmland management systems. Error bars indicate standard errors of the mean (n = 4).

Across the farmland management systems, the bacterial community composition was dominated by Actinobacteria (37%), Proteobacteria (26%), and Chloroflexi (10%) (Figure 3c, Table S2). The similarities in the relative abundance of dominant phyla and classes among farmland management systems were further investigated (Figure 3c, Table S2). The relative abundances of *r*-strategists such as Firmicutes increased 4-fold, whereas those of Myxobacteria and Bacteroidetes (phylum level) increased 0.5-fold in ISSM as compared with those of the other three systems ($p < 0.05$; Figure 3c, Table S2). The LEfSe also showed that the relative abundances of Firmicutes and Myxobacteria were the highest in ISSM ($p < 0.05$; Figures 3d and S1) than in other systems. At the class level, the relative abundances of Bacteroidia, Polyangia, Clostridia, Chloroflexi, and Bacilli were higher in ISSM ($p < 0.05$; Figure 3c, Table S2). Further comparisons revealed that the relative abundances of taxa Clostridia (belonging to the phylum Firmicutes) and Paeniclostridium, Romboutsia, and Terrisporobacter (belonging to the class Clostridia) were higher under ISSM than in other farmland management systems ($p < 0.05$; Figures 3c and S1, Table S2).

3.4. Bacterial Network and Linkage with SOM Functional Groups Depend on Management Strategy

The node and edge numbers of the bacterial co-occurrence network were high in ISSM, indicating a greater complexity of bacterial communities (Figure 4). The positive correlations between nodes were greater in the ISSM network (73%) than in the FP (16%), IP (30%), and HY (29%) networks (Figure 4a,b). ISSM increased the average clustering coefficient (*avg CC*), average closeness centrality (*avg C*), and average betweenness centrality (*avg BC*), but decreased the average path length (GD) and modularity compared to FP ($p < 0.05$; Table S5).

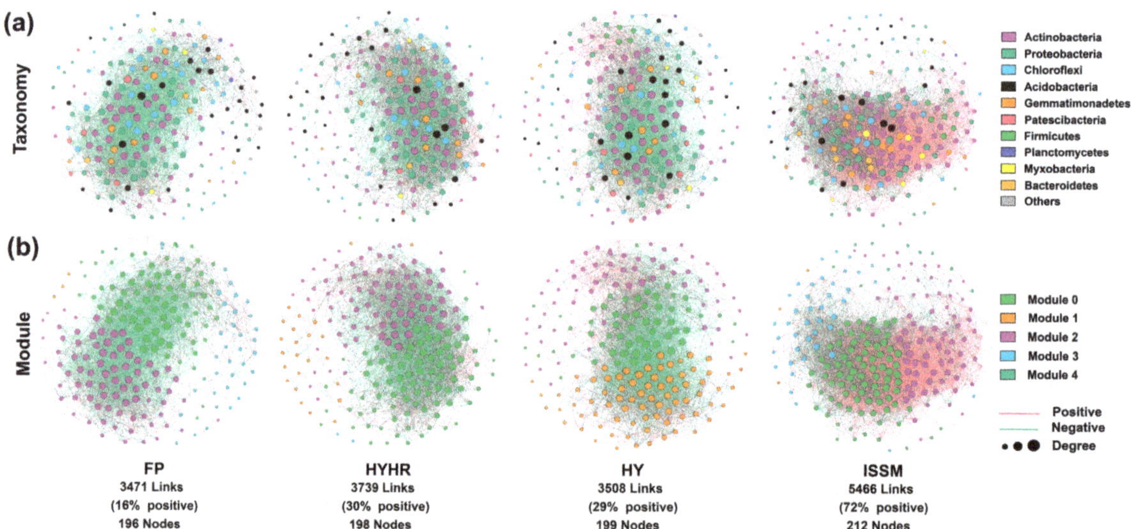

Figure 4. Co-occurrence networks of bacterial communities under four farmland management systems: FP: farmer practice; IP: improved farmer practice; HY: high-yield system; ISSM: integrated soil–crop system management. Node colors indicate phyla (**a**) and modularity classes (**b**). The size of each node is proportional to the number of degrees. For visual clarity, co-occurrence networks only show nodes with at least 10 degrees.

Co-occurrence network analysis provided evidence of the correlation between SOC chemical functional groups and bacterial community composition (Figure 5). The network pattern indicated that the correlation between labile C (O-alkyl C, Methoxyl C, and Di-O-alkyl C) and bacterial genera was 1.4 times higher under ISSM than under FP. In particular, positive correlations increased by 34% and negative correlations decreased by 55% (Figure 5a,d; Tables S6 and S7). Notably, labile C and Aryl C were positively correlated with Firmicutes (Clostridia and Bacilli) in ISSM and FP ($p < 0.05$; Figure 5, Table S7). Additionally, *Paeniclostridium*, *Romboutsia* (Clostridia), *Psychrobacillus*, and *Solibacillus* (Bacilli), positively correlated with labile C (O-alkyl C, Methoxyl C, and Di-O-alkyl C), were stronger in ISSM than in other systems ($p < 0.05$; Figure 5d, Table S2). The relative abundances of *Fluviicola* (Bacteroidetes) and *Anaeromyxobacter* (Myxobecteria) were positively correlated with Methoxyl C, and the relative abundance of *Haliangium* (Myxobecteria) was negatively correlated with Aryl C ($p < 0.05$; Figure 5d, Table S2).

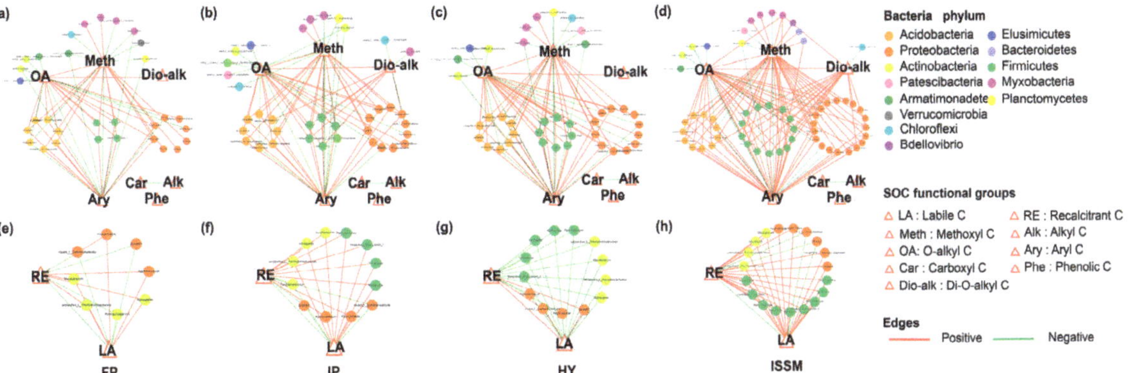

Figure 5. Network analysis revealing the associations between bacterial taxa and labile (**a–d**) and recalcitrant (**e–h**) SOC functional groups. The co-occurrence network was constructed at the genus level and colored according to the phylum classification. Seven functional groups of SOC functional groups are indicated with red triangles. Red and green lines represent strong linear connections (r > 0.6), respectively. FP: farmer practice; IP: improved farmer practice; HY: high-yield system; ISSM: integrated soil–crop system management. Labile C groups: Methoxyl C, O-alkyl C, Di-O-alkyl C, Carboxyl C. Recalcitrant C groups: Alkyl C, Aryl C, Phenolic C.

3.5. Linking SOM Quantity and Quality, Microbial Diversity, and Soil Fertility with Crop Yield

The relative abundance of Bacteroidetes, Firmicutes, and Myxobacteria were linked to soil fertility and labile C (Methoxyl C, O-alkyl C, and Di-O-alkyl C), which were positively correlated with SOC ($p < 0.05$; Figure 6a). PLS-PM analysis showed that crop yield was directly dependent on soil fertility (path coefficient = 0.58; Figure 6b), which was also supported by the Pearson correlations (Figure 6d). The SOC showed the largest effect on fertility via direct (path coefficient = 0.57) and indirect effects (path coefficient = 0.28) on the relative abundance of *r*-strategists and labile C pools (Figures 6c and S2). Soil fertility and *r*-strategists increased with SOC and labile C content (Figure 6c). There were corresponding strong positive correlations between the relative abundances of *r*-strategist bacteria and the labile C groups ($p < 0.05$; Figures 6d and S3). The effects of labile C and *r*-strategists on crop yield were indirect rather than direct (Figures 6b,c and S2). Specifically, crop yield increases were directly dependent on soil fertility. The relative abundance of *r*-strategists, SOC, and soil fertility positively correlated with crop yield ($p < 0.05$; Figure 6d). The SOC, labile C, and *r*-strategists accounted for 75% of the soil fertility index, and the soil fertility index explained 33% of the variance in total crop yield (Figure 6b).

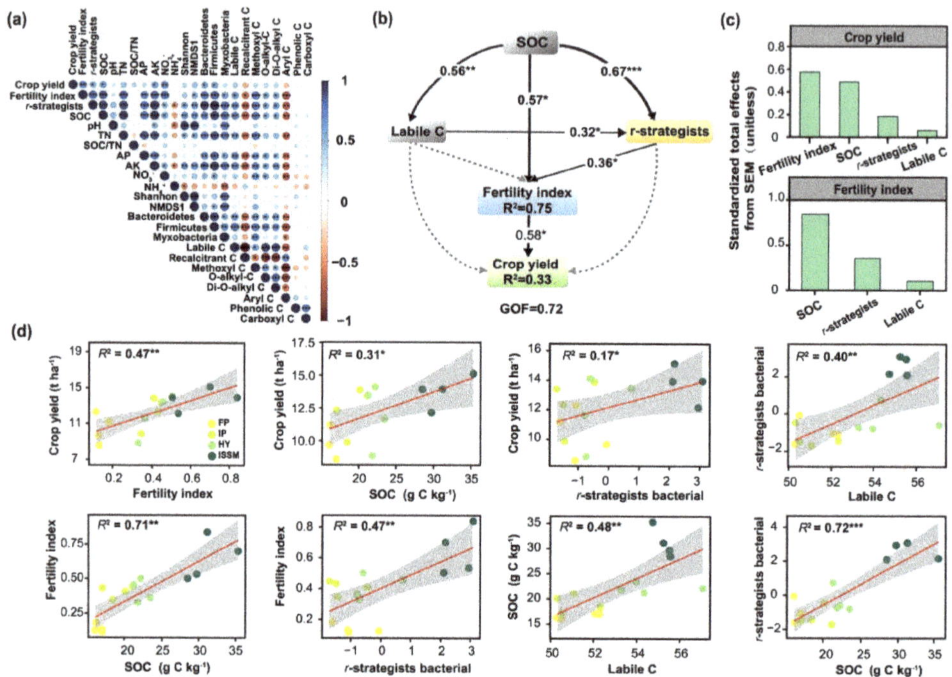

Figure 6. Correlations of crop yield, fertility index, soil properties, SOM chemical structure, and bacterial communities (**a**). Partial least-squares path (PLS−PM) analysis for crop yield, showing the relationships among soil fertility index, SOM chemical structure, and bacterial communities. GOF, goodness of fit (**b**). Black and red solid arrows indicate positive and negative associations, gray dashed arrows indicate insignificant correlations. Standardized total effects of each variable from PLS−PM (**c**). Regression of multiple factors to study the correlation between the factors (**d**). The solid line was fitted using ordinary least-squares regression and the shaded area corresponds to the 95% confidence interval. Significance codes: *** $p < 0.001$, ** $p < 0.01$, * $p < 0.05$. The fertility index was calculated based on the standardized "max–min" approach of five soil properties. The r−strategists are the first axis of the principal component analysis based on the three bacterial taxa (Support Infographic Figure S3). SOC: soil organic carbon; TN: total nitrogen; AP: available phosphorus; AK: available potassium; NH_4^+: ammonium N; NO_3^-: nitrate N.

4. Discussion

4.1. Integrated Soil–Crop Management System Increases SOM Quantity and Quality

The ISSM increased SOC by 15–87% compared to the other three systems ($p < 0.05$; Figure 1a), particularly the labile C functional groups (Methoxy C and O-alkyl C) (Figures 2c and 7). Methoxyl C and O-alkyl C are considered readily degradable components [67]. O-alkyl C groups are usually derived from carbohydrates from fresh plant material; for instance, the anomeric C is derived from cellulose and hemicellulose, whereas Methoxyl C groups are derived from fresh lignin and carbohydrates [68]. The increase in SOM, particularly labile functional C groups, may be caused by the following reasons. First, the high planting density and crop residues in ISSM can provide high amounts of labile C such as O-alkyl C [69–71]. This is consistent with the highest yield being recorded under the ISSM system (Figure 1). Second, the addition of organic fertilizers in this system resulted in a surplus of carbohydrates, including labile C [72]. Third, organic fertilizers stimulate the degradation of Aryl C groups and consequently form new Methoxyl C and O-alkyl C groups [73,74]. The high accumulation of labile C resulted in high SOC stock (Figures 1b and 2c), supporting our first hypothesis. Additionally, a lower A/O-A ratio in ISSM than in FP (Figure 2e) indicates that the SOC decomposition was slow in ISSM, resulting in more C being accumulated in the soil. These results confirm that the ISSM system reduces CO_2 emissions. Consistent with the results estimated by Cui et al. that the CO_2 emissions were reduced by 12.9% for the ISSM system-based interventions compared to FP [43]. In addition, the lower A/O-A ratio and higher Alip/Arom (Figure 2e) indicate that the ISSM system not only increases the SOC quantity, but also the quality [75]. In this study, a large amount of organic fertilizer input in ISSM systems has increased the yield and SOC content, but the problem of high nutrient losses to the environment, such as ammonia emissions, nitrate leaching, and denitrification losses, is undeniable. According to previous studies, the comprehensive results showed that ISSM system-based interventions reduced reactive nitrogen losses by 13.3–21.9% and GHG emissions by 4.6–13.2%. The yield-scaled nitrogen footprint averaged 4.6 kg reactive nitrogen loss per Mg of maize produced, compared to 6.1 kg Mg^{-1} without intervention. Similarly, yield-scaled GHG emissions were 328 kg compared to 422 kg CO_2 equivalent per Mg for maize, respectively. Farmer practice applied 300 kg^{-1} ha^{-1} of N fertilizer, while in ISSM we reduced the N fertilizer application rate to 195 kg ha^{-1} [43]. According to unpublished data from our group, the N fertilizer utilization rates of the four farm management systems were 37.3 kg kg^{-1}, 59.7 kg kg^{-1}, 39.4 kg kg^{-1}, and 62.7 kg kg^{-1}, respectively, showing that ISSM had the highest N fertilizer utilization rate and relatively reduced N losses. Therefore, ISSM systems are beneficial to C sequestration and environmental protection, and thus may serve as a sustainable farmland management practice.

4.2. Integrated Soil–Crop Management System Links Labile SOC with r-Strategists

Bacterial community composition depends on a set of environmental factors [76]. As shown in Figure 6, the relative abundances of Bacteroidetes, Firmicutes, and Myxobacteria were highly correlated with soil pH, SOC, TN, AP, and AK (Table S5). Soil pH is a vital edaphic factor affecting the diversity and composition of soil bacteria in agricultural and forest soils [77–79]. In addition to SOC and pH, other edaphic factors (e.g., TN and AK contents) strongly influenced bacterial diversity (Figure 6a); this is congruent with previous studies [80–83].

The bacterial co-occurrence network was altered by management systems. The high bacterial network complexity in ISSM (Figure 4, Table S5) is owing to the availability of a diverse range of resources (e.g., increased availability of soil carbon and nutrients, Table 2 and Figure 1a). This is in agreement with the results of studies indicating that increased soil microbial network complexity is tied to resource availability and diversity [84–86]. In general, the increased positive correlation and the reduction in the average path lengths in the ISSM co-occurrence network (Figure 4) were indicative of efficient interactions between microorganisms, promoting a denser co-occurrence network [87].

The phyla Firmicutes and Bacteroidetes prefer substrates rich in available C [31,80]. The relative abundance of Firmicutes, Myxobacteria, and Bacteroidetes increased 0.5–4 times more in ISSM than in FP (Figure 3c, Table S2). The higher relative abundance of r-strategists was due to the 25% increase in labile C functional groups (Figure 6d). In particular, *Paeniclostridium*, *Romboutsia* (Firmicutes), and *Bacteroidetes* were positively correlated with labile C (Methoxy C and O-alkyl C) (Figures 5 and 6a), confirming our second hypothesis. The *Bacillus* (Firmicutes) dominates microbial communities under labile C and available P inputs [88,89]. The relative abundance of r- (copiotrophs) strategists (e.g., *Bacteroidetes*) increased when the available substrate content in the soil was high (Figure 6a,d). Therefore, the enrichment of r-strategists under ISSM was influenced by substrate efficiency and C quality (Figure 6b). Microorganisms considered "opportunists" (r-strategists) preferentially exploit less-complex organic compounds [12,31,32]. This is also supported by the positive correlations between labile C pools and abundances of r-strategies microbial groups (Figure 6d).

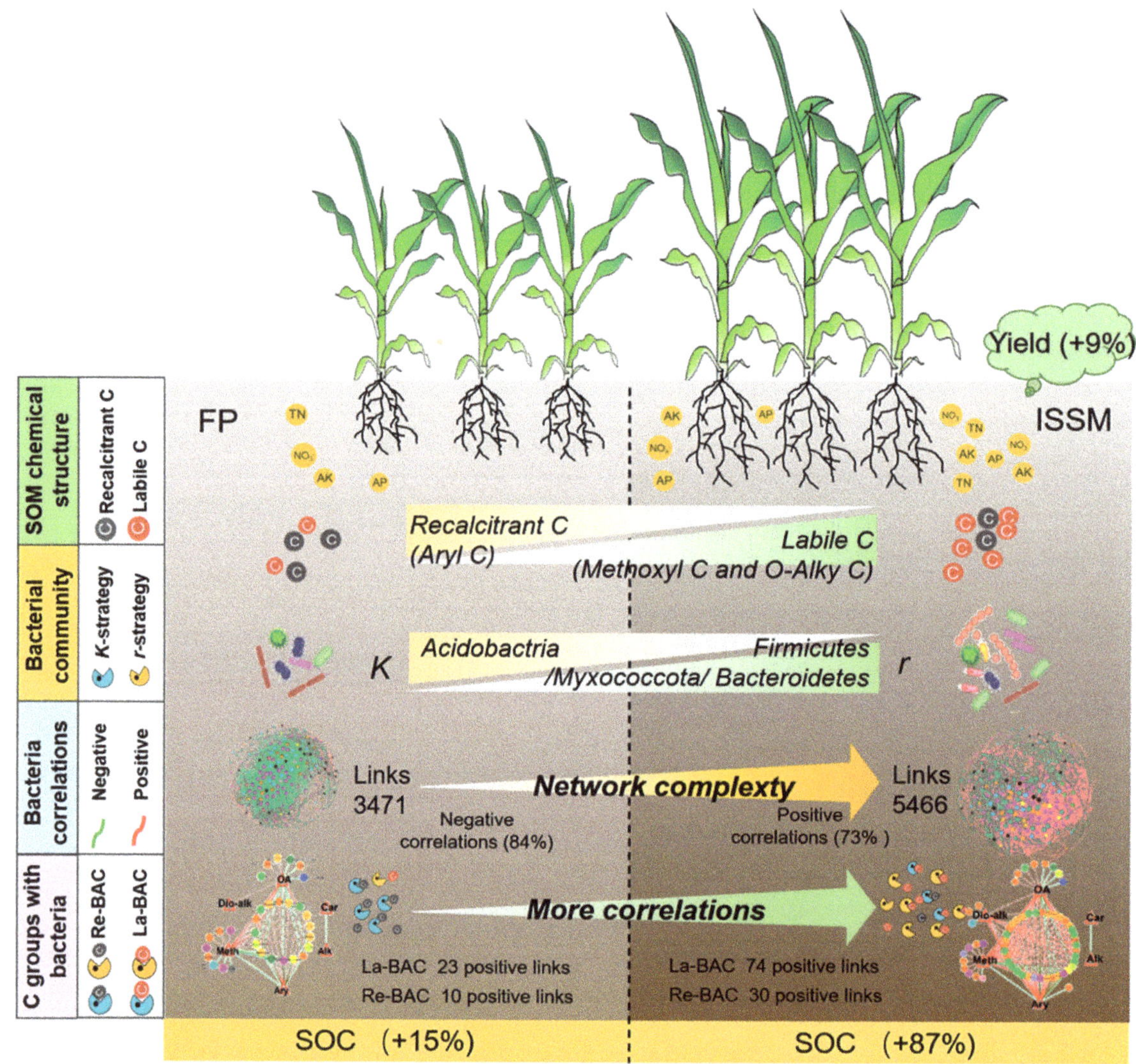

Figure 7. Concept of the effects of farmland management systems on the SOC content and compositions, and bacteria community structure and yield of maize. La: labile C; Re: recalcitrant C; BAC: bacterial community; SOM: soil organic matter; TN: total nitrogen; AP: available phosphorus; AK: available potassium; NO_3^-: nitrate N.

4.3. SOM Quantity and Quality Increased Crop Yield by Regulating Bacterial Community Composition and Enhancing Soil Fertility

Soil quality is the key to achieve high crop productivity, and soil quality is tied to SOC content and composition [90]. ISSM improved soil fertility by increasing the SOC content and abundance of *r*-strategists (Figure 6b,c), supporting the improvement of soil quality, which is more commonly associated with a transformation of bacterial community structure [91]. Therefore, the bacterial community structure is an effective biomarker for assessing the environmental impact of a wide range of agricultural practices on soil quality [92]. Specifically, significant positive correlations between SOC and soil chemistry were observed (e.g., TN, AP, AK, and NO_3^-), which is suggestive of the coupling between C sequestration and nutrient cycling, a link that explains the increase in soil fertility under ISSM (Figure 6a). Previous studies indicate that combining chemical fertilizers with organic matter such as straw and manure is an effective way to maintain soil productivity [93,94]. Congruent with these results, mixing NPK fertilizer with manure resulted in yields 25% higher than that when the NPK fertilizer was used alone. This result is directly due to enhanced soil nutrients, which leads to increased bacterial diversity [95].

Soil fertility and crop yield were significantly positively correlated (Figure 6b). Moreover, crop yield directly depended on soil fertility, and SOC and labile C indirectly affected crop yield by mediating *r*-strategists to enhance soil fertility (Figure 6b,c). Our results demonstrate that bacterial communities in carbon-rich soil primarily consist of r-strategist bacteria. These microorganisms are involved in soil nutrient cycling, which can regulate soil fertility and indirectly improve crop yields. The decomposition of plant residues and the soil's own organic matter in the soil is a biochemical process, through which carbon is returned to the atmosphere in the form of CO_2; and nitrogen, phosphorus, sulfur and trace elements are released into the soil in inorganic form for use by higher plants; some nutrients are assimilated by soil microorganisms into microbial biomass and participate in the rapid turnover process of soil microorganisms. The long-term substitution of straw and manure not only accelerated nutrient cycling, but also increased soil quality and crop yields by increasing bacterial diversity and changing the composition and chemical functional groups, which equally support our results [3,96]. Thus, reasonable agricultural management practices to improve soil fertility and C quality and further improve microbial diversity can help maintain sustainable crop production in Northeastern China [97–99].

5. Conclusions

Integrated soil–crop system management changed SOC chemical functional group levels, bacterial community structures, and their relationships with crop yield. ISSM enhanced SOC content and improved its quality by increasing labile Methoxyl C and O-alkyl C groups, resulting in a high Alip/Arom ratio. Optimization of farm management systems promotes the correlations between SOC functional groups and bacterial communities, particularly the labile Methoxyl C and O-alkyl C, with Firmicute. SOC content is one of the determinants of soil fertility, while r-strategists and labile C increase with SOC. Crop yield depends directly on soil fertility, while SOC and labile C indirectly affect crop yield by mediating *r*-strategists to enhance soil fertility. In conclusion, ISSM changed SOC quantity and quality and altered microbial community. Overall, the combined effects of these factors culminated in improved soil quality and productivity. Therefore, it is essential to optimize management strategies to maintain soil quality and long-term sustainability of agricultural production.

Supplementary Materials: The following supporting information can be downloaded at: https://www.mdpi.com/article/10.3390/agriculture13010134/s1, Table S1: Effects of farmland management systems on functional groups of SOC. Table S2: Relative abundance of soil bacteria at the phyla and class level. Table S3: Results of Adonis test for bacterial communities under four farmland management systems. Table S4: Spearman's rank correlation (R-value) between soil chemical properties and chemical structure of organic carbon and bacterial communities based on Mantel-test. Table S5: Topological properties of the empirical molecular ecological networks (MENs) of bacteria communities at four farmland management systems. Table S6: Topological properties of network of between bacterial taxa and SOC functional groups parameters of soils. Table S7: Detailed network characteristics of interactions between bacterial taxa and SOC functional groups parameters of soils. Figure S1: Linear discriminant analysis (LDA) to identify the taxa leading to community differences for bacteria. FP: farmer practice; IP: improved farmer practice; HY: high yield system; ISSM: Integrated soil-crop system management. Figure S2: Standardized effects of each variable from the partial least squares path analysis (PLS-PM). (a) and (b) represent the standardized direct effects to crop yield and fertility index, (c) and (d) represent the standardized indirect effects. The fertility index was calculated based on the standardized "Max-Min" approach of five soil properties. SOC: soil organic carbon. Labile C: Methoxyl C, O-alkyl C, Di-O-alkyl C, Carboxyl C. The relative abundances of Firmicutes,

Myxobacteria, and Bacteroidetes were chosen to *r*-strategists by principal component analysis. Figure S3: Principal component analysis (PCA) analyzing. The scores of the first PCA axis were used as *r*-strategists bacteria. The relative abundances of Firmicutes, Myxobacteria and Bacteroidetes were chosen to *r*-strategists bacteria principal components for analysis. Shaded area is 95% of the confidence interval.

Author Contributions: Conceptualization, J.T. and Q.L.; methodology, validation, and writing—original draft preparation, visualization, Q.L.; writing—review and editing, visualization, J.T, A.K. and Y.K.; resources, Z.S. and Q.G.; supervision, funding acquisition, F.Z. All authors have read and agreed to the published version of the manuscript.

Funding: This research was funded by the National Key R&D Program of China (2022YFD1901300), "Research and Application of Key Technologies for Cultivation of Arable Land Quality and Green Development of Agriculture" program (Z191100004019013), the National Natural Science Foundation of China [grant nos. 32071629], and Agriculture Carbon Neutral Account Establishment Program in Quzhou (202127). We would also like to thank the Beijing Science and Technology Program, Beijing Advanced Disciplines, and the RUDN University Strategic Academic Leadership Program.

Institutional Review Board Statement: Not applicable.

Informed Consent Statement: Not applicable.

Data Availability Statement: The data presented in this study are available on request from the corresponding author.

Conflicts of Interest: The authors declare that they have no conflicts of interest.

References

1. Schlesinger, W.H. Carbon Sequestration in Soils. *Science* **1999**, *284*, 2095. [CrossRef]
2. Zhao, Y.; Wang, M.; Hu, S.; Zhang, X.; Ouyang, Z.; Zhang, G.; Huang, B.; Zhao, S.; Wu, J.; Xie, D.; et al. Economics- and policy-driven organic carbon input enhancement dominates soil organic carbon accumulation in Chinese croplands. *Proc. Natl. Acad. Sci. USA* **2018**, *115*, 4045–4050. [CrossRef] [PubMed]
3. Lal, R. Soil Carbon Sequestration Impacts on Global Climate Change and Food Security. *Science* **2004**, *304*, 1623–1627. [CrossRef] [PubMed]
4. Paustian, K.; Lehmann, J.; Ogle, S.; Reay, D.; Robertson, G.P.; Smith, P. Climate-smart soils. *Nature* **2016**, *532*, 49–57. [CrossRef] [PubMed]
5. Nachtergaele, F.O.; Petri, M.; Biancalani, R.; Lynden, G.V.; Velthuizen, H.V.; Bloise, M. *An Information database for Land Degradation Assessment at Global Level*; Sciences and Education: Bussum, The Netherland, 2011.
6. Bossio, D.A.; Cook-Patton, S.C.; Ellis, P.W.; Fargione, J.; Sanderman, J.; Smith, P.; Wood, S.; Zomer, R.J.; von Unger, M.; Emmer, I.M.; et al. The role of soil carbon in natural climate solutions. *Nat. Sustain.* **2020**, *3*, 391–398. [CrossRef]
7. Bradford, M.; Carey, C.; Atwood, L.; Bossio, D.; Fenichel, E.; Gennet, S.; Fargione, J.; Fisher, J.; Fuller, E.; Kane, D.; et al. Soil carbon science for policy and practice. *Nat. Sustain.* **2019**, *2*, 1070–1072. [CrossRef]
8. Lal, R. Intensive Agriculture and the Soil Carbon Pool. *J. Crop Improv.* **2013**, *27*, 735–751.
9. Mikha, M.M.; Rice, C.W. Tillage and Manure Effects on Soil and Aggregate-Associated Carbon and Nitrogen. *Soil Sci. Soc. Am. J.* **2004**, *68*, 809–816. [CrossRef]
10. Wang, Y.; Hu, N.; Ge, T.; Kuzyakov, Y.; Wang, Z.-L.; Li, Z.; Tang, Z.; Chen, Y.; Wu, C.; Lou, Y. Soil aggregation regulates distributions of carbon, microbial community and enzyme activities after 23-year manure amendment. *Appl. Soil Ecol.* **2017**, *111*, 65–72. [CrossRef]
11. Zhao, Z.; Zhang, C.; Zhang, J.; Liu, C.; Wu, Q. Effects of Substituting Manure for Fertilizer on Aggregation and Aggregate Associated Carbon and Nitrogen in a Vertisol. *Agron. J.* **2019**, *111*, 368–377. [CrossRef]
12. Tian, J.; Lou, Y.; Gao, Y.; Fang, H.; Liu, S.; Xu, M.; Blagodatskaya, E.; Kuzyakov, Y. Response of soil organic matter fractions and composition of microbial community to long-term organic and mineral fertilization. *Biol. Fertil. Soils* **2017**, *53*, 523–532. [CrossRef]
13. Dutta, D.; Singh, V.K.; Upadhyay, P.K.; Meena, A.L.; Kumar, A.; Mishra, R.P.; Dwivedi, B.S.; Shukla, A.K.; Yadav, G.S.; Tewari, R.B.; et al. Long-term impact of organic and inorganic fertilizers on soil organic carbon dynamics in a rice- wheat system. *Land Degrad. Dev.* **2022**, *33*, 1862–1877. [CrossRef]
14. Jian, J.; Du, X.; Reiter, M.S.; Stewart, R.D. A meta-analysis of global cropland soil carbon changes due to cover cropping. *Soil Biol. Biochem.* **2020**, *143*, 107735. [CrossRef]
15. Wang, Y.; Wu, P.; Mei, F.; Ling, Y.; Wang, T. Does continuous straw returning keep China farmland soil organic carbon continued increase? A meta-analysis. *J. Environ. Manag.* **2021**, *288*, 112391. [CrossRef]

16. De Corato, U. Agricultural waste recycling in horticultural intensive farming systems by on-farm composting and compost-based tea application improves soil quality and plant health: A review under the perspective of a circular economy. *Sci. Total Environ.* **2020**, *738*, 139840. [CrossRef]

17. Fontaine, S.; Bardoux, G.; Abbadie, L.; Mariotti, A. Carbon input to soil may decrease soil carbon content. *Ecol. Lett.* **2010**, *7*, 314–320. [CrossRef]

18. Luo, G.; Li, L.; Friman, V.-P.; Guo, J.; Guo, S.; Shen, Q.; Ling, N. Organic amendments increase crop yields by improving microbe-mediated soil functioning of agroecosystems: A meta-analysis. *Soil Biol. Biochem.* **2018**, *124*, 105–115. [CrossRef]

19. Cai, A.; Xu, M.; Wang, B.; Zhang, W.; Liang, G.; Hou, E.; Luo, Y. Manure acts as a better fertilizer for increasing crop yields than synthetic fertilizer does by improving soil fertility. *Soil Tillage Res.* **2019**, *189*, 168–175. [CrossRef]

20. Stevenson, F.J. *Humus Chemistry: Genesis, Composition, Reactions*, 2nd ed.; Wiley: Hoboken, NJ, USA, 1994.

21. Von Lützow, M.; Kögel-Knabner, I.; Ekschmitt, K.; Flessa, H.; Guggenberger, G.; Matzner, E.; Marschner, B. SOM fractionation methods: Relevance to functional pools and to stabilization mechanisms. *Soil Biol. Biochem.* **2007**, *39*, 2183–2207. [CrossRef]

22. Jastrow, J.D.; Amonette, J.E.; Bailey, V.L. Mechanisms controlling soil carbon turnover and their potential application for enhancing carbon sequestration. *Clim. Change* **2007**, *80*, 5–23. [CrossRef]

23. Wang, Q.; HUANG, Q.; ZHANG, L.; ZHANG, J.; SHEN, Q.; RAN, W. The effects of compost in a rice–wheat cropping system on aggregate size, carbon and nitrogen content of the size–density fraction and chemical composition of soil organic matter, as shown by 13C CP NMR spectroscopy. *Soil Use Manag.* **2012**, *28*, 337–346. [CrossRef]

24. Guo, Z.; Zhang, Z.; Zhou, H.; Wang, D.; Peng, X. The effect of 34-year continuous fertilization on the SOC physical fractions and its chemical composition in a Vertisol. *Sci. Rep.* **2019**, *9*, 2505. [PubMed]

25. He, Y.T.; He, X.H.; Xu, M.G.; Zhang, W.J.; Yang, X.Y.; Huang, S.M. Long-term fertilization increases soil organic carbon and alters its chemical composition in three wheat-maize cropping sites across central and south China. *Soil Tillage Res.* **2018**, *177*, 79–87. [CrossRef]

26. Audette, Y.; Congreves, K.A.; Schneider, K.; Zaro, G.C.; Nunes, A.L.P.; Zhang, H.; Voroney, R.P. The effect of agroecosystem management on the distribution of C functional groups in soil organic matter: A review. *Biol. Fertil. Soils* **2021**, *57*, 881–894. [CrossRef]

27. Poeplau, C.; Don, A. Carbon sequestration in agricultural soils via cultivation of cover crops—A meta-analysis. *Agric. Ecosyst. Environ.* **2015**, *200*, 33–41. [CrossRef]

28. Shahbaz, M.; Kuzyakov, Y.; Sanaullah, M.; Heitkamp, F.; Zelenev, V.; Kumar, A.; Blagodatskaya, E. Microbial decomposition of soil organic matter is mediated by quality and quantity of crop residues: Mechanisms and thresholds. *Biol. Fertil. Soils* **2017**, *53*, 287–301. [CrossRef]

29. Guo, L.B.; Gifford, R.M. Soil carbon stocks and land use change: A meta-analysis. *Glob. Change Biol.* **2002**, *8*, 345–360. [CrossRef]

30. Wu, L.; Zhang, W.; Wei, W.; He, Z.; Kuzyakov, Y.; Bol, R.; Hu, R. Soil organic matter priming and carbon balance after straw addition is regulated by long-term fertilization. *Soil Biol. Biochem.* **2019**, *135*, 383–391. [CrossRef]

31. Fierer, N.; Bradford, M.A.; Jackson, R.B. Toward an ecological classification of soil bacteria. *Ecology* **2007**, *88*, 1354–1364. [CrossRef]

32. Ling, L.; Luo, Y.; Jiang, B.; Lv, J.; Meng, C.; Liao, Y.; Reid, B.J.; Ding, F.; Lu, Z.; Kuzyakov, Y.; et al. Biochar induces mineralization of soil recalcitrant components by activation of biochar responsive bacteria groups. *Soil Biol. Biochem.* **2022**, *172*, 108778. [CrossRef]

33. Allison, S.D.; Wallenstein, M.D.; Bradford, M.A. Soil-carbon response to warming dependent on microbial physiology. *Nat. Geosci.* **2010**, *3*, 336–340. [CrossRef]

34. Trivedi, D.K.; Ansari, M.W.; Dutta, T.; Singh, P.; Tuteja, N. Molecular characterization of cyclophilin A-like protein from Piriformospora indica for its potential role to abiotic stress tolerance in *E. coli.*. *BMC Res. Notes* **2013**, *6*, 555. [CrossRef]

35. Kumar, A.; Kuzyakov, Y.; Pausch, J. Maize rhizosphere priming: Field estimates using 13C natural abundance. *Plant Soil* **2016**, *409*, 87–97. [CrossRef]

36. Francioli, D.; Schulz, E.; Lentendu, G.; Wubet, T.; Buscot, F.; Reitz, T. Mineral vs. Organic Amendments: Microbial Community Structure, Activity and Abundance of Agriculturally Relevant Microbes Are Driven by Long-Term Fertilization Strategies. *Front. Microbiol.* **2016**, *7*, 1446. [CrossRef]

37. Semenov, M.V.; Krasnov, G.S.; Semenov, V.M.; Ksenofontova, N.; Zinyakova, N.B.; van Bruggen, A.H.C. Does fresh farmyard manure introduce surviving microbes into soil or activate soil-borne microbiota? *J. Environ. Manag.* **2021**, *294*, 113018. [CrossRef] [PubMed]

38. Strap, J.L. Actinobacteria–Plant Interactions: A Boon to Agriculture. In *Bacteria in Agrobiology: Plant Growth Responses*; Maheshwari, D.K., Ed.; Springer: Berlin/Heidelberg, Germany, 2011; pp. 285–307.

39. Bryanin, S.; Abramova, E.; Makoto, K. Fire-derived charcoal might promote fine root decomposition in boreal forests. *Soil Biol. Biochem.* **2018**, *116*, 1–3. [CrossRef]

40. Li, X.; Chen, Q.L.; He, C.; Shi, Q.; Chen, S.C.; Reid, B.J.; Zhu, Y.G.; Sun, G.X. Organic Carbon Amendments Affect the Chemodiversity of Soil Dissolved Organic Matter and Its Associations with Soil Microbial Communities. *Environ. Sci. Technol.* **2019**, *53*, 50–59. [CrossRef]

41. Chen, X.P.; Cui, Z.L.; Vitousek, P.M.; Cassman, K.G.; Matson, P.A.; Bai, J.S.; Meng, Q.F.; Hou, P.; Yue, S.C.; Romheld, V. Integrated soil–crop system management for food security. *Proc. Natl. Acad. Sci. USA* **2011**, *108*, 6399–6404. [CrossRef]

42. Chen, X.; Cui, Z.; Fan, M.; Vitousek, P.; Zhao, M.; Ma, W.; Wang, Z.; Zhang, W.; Yan, X.; Yang, J.; et al. Producing more grain with lower environmental costs. *Nature* **2014**, *514*, 486–489. [CrossRef]

43. Cui, Z.; Zhang, H.; Chen, X.; Zhang, C.; Ma, W.; Huang, C.; Zhang, W.; Mi, G.; Miao, Y.; Li, X.; et al. Pursuing sustainable productivity with millions of smallholder farmers. *Nature* **2018**, *555*, 363–366. [CrossRef]

44. Li, B.; Liu, Z.; Huang, F.; Yang, X.; Liu, Z.; Wan, W.; Wang, J.; Xu, Y.; Li, Z.; Ren, T. Ensuring National Food Security by Strengthening High-productivity Black Soil Granary in Northeast China. *Bull. Chin. Acad. Sci.* **2021**, *36*, 1184–1193.

45. Yan, X.; CAI, Z.; WANG, S.; SMITH, P. Direct measurement of soil organic carbon content change in the croplands of China. *Glob. Change Biol.* **2011**, *17*, 1487–1496. [CrossRef]

46. Liang, A.; Yang, X.; Zhang, X.; Mclaughlin, N.; Yan, S.; Li, W. Soil organic carbon changes in particle-size fractions following cultivation of Black soils in China. *Soil Tillage Res.* **2009**, *105*, 21–26. [CrossRef]

47. Liu, X.; Han, X.; Song, C.; Herbert, S.J.; Xing, B. Soil Organic Carbon Dynamics in Black Soils of China Under Different Agricultural Management Systems. *Commun. Soil Sci. Plant Anal.* **2003**, *34*, 973–984. [CrossRef]

48. Liu, B.; Xia, H.; Jiang, C.; Riaz, M.; Yang, L.; Chen, Y.; Fan, X.; Xia, X. 14-year applications of chemical fertilizers and crop straw effects on soil labile organic carbon fractions, enzyme activities and microbial community in rice-wheat rotation of middle China. *Sci. Total Environ.* **2022**, *841*, 156608. [CrossRef] [PubMed]

49. Malik, A.A.; Martiny, J.B.H.; Brodie, E.L.; Martiny, A.C.; Treseder, K.K.; Allison, S.D. Defining trait-based microbial strategies with consequences for soil carbon cycling under climate change. *ISME J.* **2020**, *14*, 1–9. [CrossRef] [PubMed]

50. Kuzyakov, Y.; Blagodatskaya, E. Microbial hotspots and hot moments in soil: Concept & review. *Soil Biol. Biochem.* **2015**, *83*, 184–199. [CrossRef]

51. Hanway, J.J.; Heidal, H. *Soil Analysis Methods Asused in Iowa State College Soil Testing Laboratory*; FAO: Rome, Italy, 1952.

52. Olsen, S.R. *Estimation of Available Phosphorus in Soils by Extraction with Sodium Bicarbonate*; US Department of Agriculture: Washington, DC, USA, 1954.

53. Novara, A.; Poma, I.; Sarno, M.; Venezia, G.; Gristina, L. Long-Term Durum Wheat-Based Cropping Systems Result in the Rapid Saturation of Soil Carbon in the Mediterranean Semi-arid Environment. *Land Degrad. Dev.* **2016**, *27*, 612–619. [CrossRef]

54. Delgado-Baquerizo, M.; Reich, P.B.; Trivedi, C.; Eldridge, D.J.; Abades, S.; Alfaro, F.D.; Bastida, F.; Berhe, A.A.; Cutler, N.A.; Gallardo, A.; et al. Multiple elements of soil biodiversity drive ecosystem functions across biomes. *Nat. Ecol. Evol.* **2020**, *4*, 210–220. [CrossRef]

55. Ma, L.; Ju, Z.; Fang, Y.; Vancov, T.; Gao, Q.; Wu, D.; Zhang, A.; Wang, Y.; Hu, C.; Wu, W.; et al. Soil warming and nitrogen addition facilitates lignin and microbial residues accrual in temperate agroecosystems. *Soil Biol. Biochem.* **2022**, *170*, 108693. [CrossRef]

56. Bonanomi, G.; Incerti, G.; Giannino, F.; Mingo, A.; Lanzotti, V.; Mazzoleni, S. Litter quality assessed by solid state 13C NMR spectroscopy predicts decay rate better than C/N and Lignin/N ratios. *Soil Biol. Biochem.* **2013**, *56*, 40–48. [CrossRef]

57. Baldock, J.A.; Oades, J.M.; Waters, A.G.; Peng, X.; Vassallo, A.M.; Wilson, M.A. Aspects of the chemical structure of soil organic materials as revealed by solid-state13C NMR spectroscopy. *Biogeochemistry* **1992**, *16*, 1–42. [CrossRef]

58. Coxall, H.K.; Wilson, P.A.; Pälike, H.; Lear, C.H.; Backman, J. Rapid stepwise onset of Antarctic glaciation and deeper calcite compensation in the Pacific Ocean. *Nature* **2005**, *433*, 53–57. [CrossRef] [PubMed]

59. Li, Y.; Jiang, P.; Chang, S.X.; Wu, J.; Lin, L. Organic mulch and fertilization affect soil carbon pools and forms under intensively managed bamboo (*Phyllostachys praecox*) forests in southeast China. *J. Soils Sediments* **2010**, *10*, 739–747. [CrossRef]

60. Boeni, M.; Bayer, C.; Dieckow, J.; Conceição, P.C.; Dick, D.P.; Knicker, H.; Salton, J.C.; Macedo, M.C.M. Organic matter composition in density fractions of Cerrado Ferralsols as revealed by CPMAS 13C NMR: Influence of pastureland, cropland and integrated crop-livestock. *Agric. Ecosyst. Environ.* **2014**, *190*, 80–86. [CrossRef]

61. Edgar, R.C. UPARSE: Highly accurate OTU sequences from microbial amplicon reads. *Nat. Methods* **2013**, *10*, 996–998. [CrossRef]

62. Cole, J.; Crowle, S.; Austwick, G.; Henderson Slater, D. Exploratory findings with virtual reality for phantom limb pain; from stump motion to agency and analgesia. *Disabil. Rehabil.* **2009**, *31*, 846–854. [CrossRef]

63. Segata, N.; Izard, J.; Waldron, L.; Gevers, D.; Miropolsky, L.; Huttenhower, G.C. Metagenomic biomarker discovery and explanation. *Genome Biol.* **2011**, *12*, R60. [CrossRef]

64. Krzywinski, M.; Schein, J.; Birol, I.; Connors, J.; Gascoyne, R.; Horsman, D.; Jones, S.J.; Marra, M.A. Circos: An information aesthetic for comparative genomics. *Genome Res.* **2009**, *19*, 1639–1645. [CrossRef]

65. Benjamini, Y.; Hochberg, Y. Controlling the false discovery rate: A new and powerful approach to multiple testing. *J. R. Stat. Soc. Ser. B Methodol.* **1995**, *57*, 289–300.

66. Benjamini, Y.; Hochberg, Y. On the Adaptive Control of the False Discovery Rate in Multiple Testing with Independent Statistics. *J. Educ. Behav. Stat.* **2000**, *25*, 60–83. [CrossRef]

67. Zhang, Y.; Yao, S.; Mao, J.; Olk, D.C.; Cao, X.; Zhang, B. Chemical composition of organic matter in a deep soil changed with a positive priming effect due to glucose addition as investigated by C-13 NMR spectroscopy. *Soil Biol. Biochem.* **2015**, *85*, 137–144. [CrossRef]

68. Zhou, Z.; Chen, N.; Cao, X.; Chua, T.; Mao, J.; Mandel, R.D.; Bettis, E.A.; Thompson, M.L. Composition of clay-fraction organic matter in Holocene paleosols revealed by advanced solid-state NMR spectroscopy. *Geoderma* **2014**, *223–225*, 54–61. [CrossRef]

69. Arshad, M.A.; Soon, Y.K.; Ripmeester, J.A. Quality of soil organic matter and C storage as influenced by cropping systems in northwestern Alberta, Canada. *Nutr. Cycl. Agroecosystems* **2011**, *89*, 71–79. [CrossRef]

70. Shrestha, B.M.; Singh, B.R.; Forte, C.; Certini, G. Long-term effects of tillage, nutrient application and crop rotation on soil organic matter quality assessed by NMR spectroscopy. *Soil Use Manag.* **2015**, *31*, 358–366. [CrossRef]

71. Soon, Y.K.; Arshad, M.A.; Haq, A.; Lupwayi, N. The influence of 12 years of tillage and crop rotation on total and labile organic carbon in a sandy loam soil. *Soil Tillage Res.* **2007**, *95*, 38–46. [CrossRef]
72. Li, D.; Chen, L.; Xu, J.; Ma, L.; Olk, D.C.; Zhao, B.; Zhang, J.; Xin, X. Chemical nature of soil organic carbon under different long-term fertilization regimes is coupled with changes in the bacterial community composition in a Calcaric Fluvisol. *Biol. Fertil. Soils* **2018**, *54*, 999–1012. [CrossRef]
73. De Marco, A.; Spaccini, R.; Vittozzi, P.; Esposito, F.; Berg, B.; Virzo De Santo, A. Decomposition of black locust and black pine leaf litter in two coeval forest stands on Mount Vesuvius and dynamics of organic components assessed through proximate analysis and NMR spectroscopy. *Soil Biol. Biochem.* **2012**, *51*, 1–15. [CrossRef]
74. Li, Y.; Chen, N.; Harmon, M.E.; Li, Y.; Cao, X.; Chappell, M.A.; Mao, J. Plant Species Rather Than Climate Greatly Alters the Temporal Pattern of Litter Chemical Composition During Long-Term Decomposition. *Sci. Rep.* **2015**, *5*, 15783. [CrossRef]
75. Demyan, M.S.; Rasche, F.; Schulz, E.; Breulmann, M.; Müller, T.; Cadisch, G. Use of specific peaks obtained by diffuse reflectance Fourier transform mid-infrared spectroscopy to study the composition of organic matter in a Haplic Chernozem. *Eur. J. Soil Sci.* **2012**, *63*, 189–199. [CrossRef]
76. Garland, G.; Edlinger, A.; Banerjee, S.; Degrune, F.; García-Palacios, P.; Pescador, D.S.; Herzog, C.; Romdhane, S.; Saghai, A.; Spor, A.; et al. Crop cover is more important than rotational diversity for soil multifunctionality and cereal yields in European cropping systems. *Nat. Food* **2021**, *2*, 28–37. [CrossRef]
77. Rousk, J.; Bååth, E.; Brookes, P.C.; Lauber, C.L.; Lozupone, C.; Caporaso, J.G.; Knight, R.; Fierer, N. Soil bacterial and fungal communities across a pH gradient in an arable soil. *ISME J.* **2010**, *4*, 1340–1351. [CrossRef] [PubMed]
78. Wessén, E.; Hallin, S.; Philippot, L. Differential responses of bacterial and archaeal groups at high taxonomical ranks to soil management. *Soil Biol. Biochem.* **2010**, *42*, 1759–1765. [CrossRef]
79. Fierer, N.; Jackson, R.B. The diversity and biogeography of soil bacterial communities. *Proc. Natl. Acad. Sci. USA* **2006**, *103*, 626–631. [CrossRef] [PubMed]
80. Ramirez, K.S.; Lauber, C.L.; Knight, R.; Bradford, M.A.; Fierer, N. Consistent effects of nitrogen fertilization on soil bacterial communities in contrasting systems. *Ecology* **2010**, *91*, 3463–3470. [CrossRef]
81. Yuan, H.; Ge, T.; Wu, X.; Liu, S.; Tong, C.; Qin, H.; Wu, M.; Wei, W.; Wu, J. Long-term field fertilization alters the diversity of autotrophic bacteria based on the ribulose-1,5-biphosphate carboxylase/oxygenase (RubisCO) large-subunit genes in paddy soil. *Appl. Microbiol. Biotechnol.* **2012**, *95*, 1061–1071. [CrossRef]
82. Bergkemper, F.; Schöler, A.; Engel, M.; Lang, F.; Krüger, J.; Schloter, M.; Schulz, S. Phosphorus depletion in forest soils shapes bacterial communities towards phosphorus recycling systems. *Environ. Microbiol.* **2016**, *18*, 1988–2000. [CrossRef]
83. De Vries, F.T.; Bloem, J.; Quirk, H.; Stevens, C.J.; Bol, R.; Bardgett, R.D. Extensive management promotes plant and microbial nitrogen retention in temperate grassland. *PLoS ONE* **2012**, *7*, e51201. [CrossRef]
84. Barberán, A.; Bates, S.T.; Casamayor, E.O.; Fierer, N. Using network analysis to explore co-occurrence patterns in soil microbial communities. *ISME J.* **2012**, *6*, 343–351. [CrossRef]
85. Banerjee, S.; Walder, F.; Büchi, L.; Meyer, M.; Held, A.Y.; Gattinger, A.; Keller, T.; Charles, R.; van der Heijden, M.G.A. Agricultural intensification reduces microbial network complexity and the abundance of keystone taxa in roots. *ISME J.* **2019**, *13*, 1722–1736. [CrossRef]
86. Banerjee, S.; Kirkby, C.A.; Schmutter, D.; Bissett, A.; Kirkegaard, J.A.; Richardson, A.E. Network analysis reveals functional redundancy and keystone taxa amongst bacterial and fungal communities during organic matter decomposition in an arable soil. *Soil Biol. Biochem.* **2016**, *97*, 188–198. [CrossRef]
87. Yuan, M.M.; Guo, X.; Wu, L.; Zhang, Y.; Zhou, J. Climate warming enhances microbial network complexity and stability. *Nat. Clim. Change* **2021**, *11*, 343–348. [CrossRef]
88. Ramirez, K.S.; Craine, J.M.; Fierer, N. Consistent effects of nitrogen amendments on soil microbial communities and processes across biomes. *Glob. Change Biol.* **2012**, *18*, 1918–1927. [CrossRef]
89. Jing, Z.; Chen, R.; Wei, S.; Feng, Y.; Zhang, J.; Lin, X. Response and feedback of C mineralization to P availability driven by soil microorganisms. *Soil Biol. Biochem.* **2017**, *105*, 111–120. [CrossRef]
90. Lal, R.; Negassa, W.; Lorenz, K. Carbon sequestration in soil. *Curr. Opin. Environ. Sustain.* **2015**, *15*, 79–86. [CrossRef]
91. Kennedy, A.C.; Smith, K.L. Soil microbial diversity and the sustainability of agricultural soils. *Plant Soil* **1995**, *170*, 75–86. [CrossRef]
92. Sharma, S.K.; Ramesh, A.; Sharma, M.P.; Joshi, O.P.; Karlen, D.L. Microbial Community Structure and Diversity as Indicators for Evaluating Soil Quality. In *Biodiversity, Biofuels, Agroforestry and Conservation Agriculture*; Springer: Berlin/Heidelberg, Germany, 2010.
93. Steiner, C.; Teixeira, W.G.; Lehmann, J.; Nehls, T.; de Macêdo, J.L.V.; Blum, W.E.H.; Zech, W. Long term effects of manure, charcoal and mineral fertilization on crop production and fertility on a highly weathered Central Amazonian upland soil. *Plant Soil* **2007**, *291*, 275–290. [CrossRef]
94. Verma, S.; Sharma, P.K. Long-term effects of organics, fertilizers and cropping systems on soil physical productivity evaluated using a single value index (NLWR). *Soil Tillage Res.* **2008**, *98*, 1–10. [CrossRef]
95. Sun, R.; Zhang, X.-X.; Guo, X.; Wang, D.; Chu, H. Bacterial diversity in soils subjected to long-term chemical fertilization can be more stably maintained with the addition of livestock manure than wheat straw. *Soil Biol. Biochem.* **2015**, *88*, 9–18. [CrossRef]

96. Liu, H.; Du, X.; Li, Y.; Han, X.; Li, B.; Zhang, X.; Li, Q.; Liang, W. Organic substitutions improve soil quality and maize yield through increasing soil microbial diversity. *J. Clean. Prod.* **2022**, *347*, 131323. [CrossRef]
97. Mazzilli, S.R.; Kemanian, A.R.; Ernst, O.R.; Jackson, R.B.; Piñeiro, G. Priming of soil organic carbon decomposition induced by corn compared to soybean crops. *Soil Biol. Biochem.* **2014**, *75*, 273–281. [CrossRef]
98. Xie, H.; Li, J.; Zhu, P.; Peng, C.; Wang, J.; He, H.; Zhang, X. Long-term manure amendments enhance neutral sugar accumulation in bulk soil and particulate organic matter in a Mollisol. *Soil Biol. Biochem.* **2014**, *78*, 45–53. [CrossRef]
99. Lin, Y.; Ye, G.; Kuzyakov, Y.; Liu, D.; Fan, J.; Ding, W. Long-term manure application increases soil organic matter and aggregation, and alters microbial community structure and keystone taxa. *Soil Biol. Biochem.* **2019**, *134*, 187–196. [CrossRef]

Article

Spatiotemporal Patterns and Influencing Factors of Agriculture Methane Emissions in China

Guofeng Wang [1], Pu Liu [1], Jinmiao Hu [1] and Fan Zhang [2,*]

1   Faculty of International Trade, Shanxi University of Finance and Economics, Taiyuan 030006, China
2   Institute of Geographical Sciences and Natural Resources Research, Chinese Academy of Sciences, Beijing 100101, China
*   Correspondence: zhangf.ccap@igsnrr.ac.cn

Academic Editors: Yinglong Chen, Masanori Saito and Etelvino Henrique Novotny

Received: 29 August 2022
Accepted: 24 September 2022
Published: 29 September 2022

Abstract: Explaining the methane emission pattern of Chinese agriculture and the influencing factors of its spatiotemporal differentiation is of great theoretical and practical significance for carbon neutrality. This paper uses the IPCC coefficient method to measure and analyze the spatial and temporal differentiation characteristics of agricultural methane emission, clarify the dynamic evolution trend of the kernel density function, and reveal the key influencing factors of agricultural methane emission with geographical detectors. The results show that China's agricultural methane emissions showed a first increasing and then declining trend. Agricultural methane emissions decreased from 21.4587 million tons to 17.6864 million tons, with an upward trend from 2000 to 2005, a significant decline in 2006, a slow change from 2007 to 2015, and a significant decline from 2015 to 2019. In addition, the emissions pattern of the three major grain functional areas is characteristic; in 2019, agricultural methane emissions from main producing area, main sales area, and balance area were 10.8406 million tons, 1.2471 million tons, and 5.599 million tons, respectively. The main grain producing area is the main area of methane emissions, and the emission pattern will not change in the short term. The variability of grain functional areas is the decisive factor for the difference in agricultural methane emissions. The state of industrial structure is the key influencing factor for adjusting the spatial distribution—the explanatory power of the industrial structure to the main producing areas reached 0.549; the level of agricultural development is the most core influencing factor of the spatial pattern of the main grain sales area—the explanatory power reached 0.292; and the level of industrialization and the industrial structure are the core influencing factors of the spatial pattern of the balance area—the explanatory power reached 0.545 and 0.479, respectively.

Keywords: agriculture; methane emission; spatiotemporal pattern; kernel density; influencing factors

1. Introduction

Climate change has become the biggest threat to global sustainable development [1–4]. The Paris Agreement stipulates that the state parties should keep the global average warming to 2 degrees Celsius higher than the pre-industrial revolution level and strive to limit it to 1.5 degrees Celsius [5]. The "climate critical points" of 1.5 and 2 degrees Celsius are the key time points for catastrophic climate events but are also related to the threat of human survival, which may produce a chain reaction once exceeded [6]. In addition, frequent occurrence of global extreme climate caused by climate change has brought significant economic and social losses to human beings [7]. How to mitigate and adapt to climate change has become a major issue for countries all over the world. Carbon peaking and carbon neutrality is a solemn commitment of China to address global climate change, and a major declaration of the transformation of social and economic development mode [8]. To achieve carbon peaking by 2030 and strive to carbon neutrality by 2060 is not only an essential part of China's high-quality growth but also a reform requirement for major adjustments in different industries [9].

Methane is the second largest greenhouse gas responsible for climate change after carbon dioxide [10]. Strengthening methane emissions reduction has become a necessary item in the 21st century [11]. In 2020, global methane emissions were 570 million tons, of which human activities produced 340 million tons. IPCC pointed out in Working Group I of the Sixth Assessment Report that within 20 years after emission, the greenhouse effect of 1 ton of methane is comparable to 84 tons of carbon dioxide and its warming effect is still 28 times that of carbon dioxide even after 100 years [12]. In addition, since it is easier to reduce methane than carbon dioxide, the International Energy Agency (IEA) notes that 75% of methane leaks in the global oil and gas industry can be controlled with existing technologies, and 50% of methane emissions are reduced at net zero cost. [13]. Methane has a shorter lifetime in the atmosphere than carbon dioxide and can be a priority for emissions reduction. Emission reductions of non-carbon dioxide greenhouse gases such as methane are essential for the goal of controlling global warming to 1.5 °C by the end of this century [14–16].

As agriculture is the main source of methane emissions, it is of great significance to clarify its spatial distribution and influencing factors [17]. Methane emissions from agriculture account for about 1/5 of the total global emissions, and the main sources of agricultural methane emissions are farming and animal husbandry. The carbon emissions produced by the planting industry are mainly generated by paddy planting, while the carbon emissions produced by the breeding industry are mainly generated by the intestinal fermentation and manure management of livestock and poultry [18].

Scholars from all over the world have studied agricultural methane emissions from different perspectives. Some scholars have estimated the intestinal methane emissions of livestock and poultry in the Southern African Development Community (SADC) [19]. Some scholars have studied methane emissions from wetland rice fields [20]. In addition, some researchers have compared the intestinal methane emissions of livestock and poultry in Germany in the late 19th century with the current situation [21]. As a major agricultural country, China's methane emissions from agricultural activities account for 50.15% of the country's total emissions. Agricultural methane emission patterns and influencing factors have become a hot topic in academic research [22–27]. After estimating the methane emissions of various regions in China, it is indicated that the total methane emissions from agricultural interaction in 2018 were 18.22 million tons, among which the intestinal fermentation emissions from livestock and poultry were the largest, accounting for 50.69%, the emissions from paddy were 35.17%, and the emissions from livestock and poultry management were 14.14% [28]. By measuring methane and nitrous oxide emissions from 1980 to 2018, Li Yang pointed out that the methane missions generated from agriculture increased from 0.56×10^9 CO_2-eq to 0.73×10^9 CO_2-eq, which were mainly affected by efficiency factor, structure factor, and population size factor [29]. In addition, scholars' research focused on the provincial level, and calculated the changes in Jiangxi, Guangdong, Heilongjiang, and other provinces. Some scholars have also measured methane emissions from lake farms and their response to ecological restoration [30], which provided a solid foundation to study agricultural methane emissions. However, at present, there are relatively few estimates based on the national level, and there is a lack of comparative studies on the differences among main grain producing area, main sales area, and balance area; thus, relevant influencing factors need to be further clarified. In this paper, the spatial and temporal patterns of methane emission in China were analyzed by using the IPCC method and the calculation method of the Provincial Greenhouse Gas Inventory Guide. The influencing factors were calculated by using the analysis method of the geographic detector, and the emission reduction countermeasures and suggestions were put forth to provide some references for methane emission reduction in China.

2. Materials and Methods

2.1. Measurement Method of Agricultural Methane Emission

The CH_4 emissions of the national agricultural system mainly come from paddy fields and livestock and poultry manure management and intestinal fermentation [28]. Among them, the CH_4 emissions from paddy fields generally follow the basic methods and requirements determined by the IPCC guidelines, and the formula is calculated as:

$$E_{CH_4 rice} = \sum EF_i \times AD_i \tag{1}$$

where $E_{CH_4 rice}$ represents CH_4 emissions from paddy fields ($\times 10^4$ t), EF_i represents methane emission factors from paddy fields (kg/hm^2), and AD_i represents the sown area of this type of methane emission factor ($\times 10^3$ hm^2),The methane emission factors of paddy fields are listed in Table 1.

Table 1. Methane emission factors from paddy fields.

	North China	East China	Central South	Southwest	Northeast	Northwest
Single-cropping rice	234.0	215.5	236.7	156.2	168.0	231.2
Double-cropping early rice	—	211.4	241.0	156.2	—	—
Double-cropping late rice	—	224.0	273.2	171.7	—	—

The calculation formula of CH_4 generated by livestock and poultry manure management is calculated as follows:

$$E_{CH_4,manure,i} = EF_{CH_4,manure,i} \times AP_i \times 10^{-7} \tag{2}$$

where $E_{CH_4,manure,i}$ refers to the amount of methane produced by manure management of species i ($\times 10^4$ t), $EF_{CH_4,manure,i}$ refers to manure management methane emission factor of species i, and AP_i refers to the number of species i, the methane emission factors for manure management are listed in Table 2.

Table 2. Methane emission factors from manure management.

	North China	Northeast	East China	Central South	Southwest	Northwest
Dairy cow	7.46	2.23	8.33	8.45	6.51	5.93
Non-dairy cow	2.82	1.02	3.31	4.72	3.21	1.86
Sheep	0.15	0.15	0.26	0.34	0.48	0.28
Goat	0.17	0.16	0.28	0.31	0.53	0.32
Pig	3.12	1.12	5.08	5.85	4.18	1.38
Poultry	0.01	0.01	0.02	0.02	0.02	0.01
Horse	1.09	1.09	1.64	1.64	1.64	1.09
Donkey/mule	0.60	0.60	0.90	0.90	0.90	0.60
Camel	1.28	1.28	1.92	1.92	1.92	1.28

The calculation formula of CH_4 produced by intestinal fermentation of livestock and poultry is calculated as follows:

$$E_{CH_4,enteric,i} = \sum EF_{CH_4,enteric,i,j} \times AP_i \times R_j \times 10^{-7} \tag{3}$$

where $E_{CH_4,enteric,i}$ is the amount of methane produced by intestinal fermentation of species i ($\times 10^4$ t), $EF_{CH_4,enteric,i,j}$ is methane emission factor from intestinal fermentation of livestock and poultry of species I, AP_i is the number of species I, and R_j is the breeding proportion of this livestock and poultry, The methane emission factor from enteric fermentation of livestock and poultry are listed in Table 3.

Table 3. Methane emission factor from intestinal fermentation of livestock and poultry.

Feeding Way	Dairy Cow	Cow	Sheep	Pig	Horse	Donkey	Mule	Camel
Scale feeding	88.1	52.9	8.9	1	18	10	10	46
Farmers free-ranging	89.3	67.9	9.4	1	18	10	10	46

2.2. Kernel Density Function

The kernel density function, due to its non-parametric estimation power of the probability density, can be used to measure the distribution form of random variables, with the expression of:

$$f(x) = \frac{1}{nh}\sum_{i=1}^{n} K(\frac{x - x_i}{h}) \tag{4}$$

where n represents the number of the sample, h represents bandwidth, $h = 0.9SN^{4/5}$ (N represents the number of the sample and S represents sample standard deviation). $K(\frac{x-x_i}{h})$ represents kernel density function, and the Epanechnikov kernel density form is used. The dynamic evolution of methane emission zones can be reflected by the distribution interval, morphology, and kurtosis extension of kernel density function. If the function presents a "single peak" as a whole, there are no multiple equilibrium states; if the function appears a "double peak" or "multi-peak" state, there are two or more equilibrium points.

2.3. Geographic Detector

To deeply explore the causes of regional differences in agricultural methane emissions, geographic detector analysis is used in this study. Geographic detectors can not only test the spatial differentiation of the delay variables but also excavate the influencing factors of spatial differentiation, explain the decisive role of influencing factors, and provide panoramic display for the exploration of the spatial differences. In this study, the factor detection method is adopted to determine the fundamental factor of spatial differentiation of agricultural methane emissions through the determinant force index q, in which Y represents the agricultural methane emission, $X = \{X_m\}$ represents influencing factors, $(m = 1, 2, \ldots L;)$, L refers to the number of partitions of factor X, and X_m refers to the different partitions of factor X. The determinant force index q of factor X on Y is as follows:

$$q = 1 - \frac{\sum_{m}^{L} N_m \sigma_m^2}{N\sigma^2} = 1 - \frac{SSW}{SST}$$
$$SSW = \sum_{m}^{L} N_m \sigma_m^2 \tag{5}$$
$$SST = N\sigma^2$$

where N is the number of provinces in the study, N_m is the number of provinces contained in the m-th partition of factor X, σ^2 is the variance of region Y, σ_m^2 is the variance of the driver X in the m subdomain, and SSW and SST represent the sum of variance in each grain function and the total variance in the whole region, respectively. In general, the larger the q-value of factor X, the stronger the factor of driving force for the spatial analysis. When the driving factors X and Y have driving effects on each other, the sum of the variances within regions are usually less than the sum of the variances between regions. In order to compare whether the cumulative variance of different grain production regions is significantly different from the overall variance of the whole region, the F-statistic test is introduced in this paper.

2.4. Data

In this paper, the paddy fields area, livestock and poultry quantity, and breeding proportion are from the China Statistical Yearbook (2000–2019), State Administration of Grain, China Agriculture Yearbook, and China Animal Husbandry and Veterinary Yearbook (2000–2019). Methane emission factors in different paddy fields, from livestock and poultry manure management in different regions, and from livestock and poultry intestinal methane emission factors under different feeding methods are from the Provincial Guidelines for Greenhouse Gas (National Development and Reform Commission, 2011), and some missing data are supplemented by a linear interpolation method.

3. Results

3.1. Measurement of Agricultural Methane Emission in China

3.1.1. Overall Agricultural Methane Sequential Variation

Agricultural methane emissions showed a trend of first increasing and then decreasing. From 2000 to 2019 (Figure 1), agricultural methane emissions decreased from 21.46 million tons to 17.69 million tons, with an average annual drop of 0.88%. Overall, China's green development in agriculture has been significantly improved. From 2000 to 2005, agricultural methane emissions continued to increase, with a significant decline in 2006.

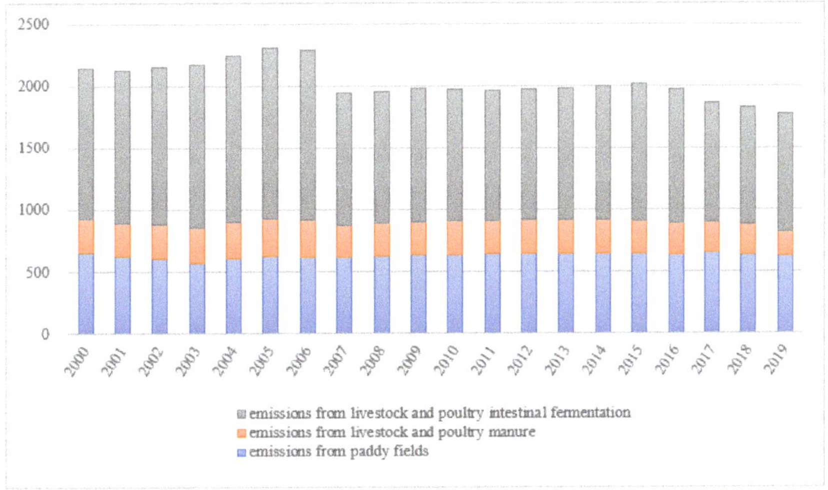

Figure 1. Agricultural methane emissions in China, 2000–2019.

From 2006 to 2015, there was a slow change period, with a slight increase in methane emissions, and since 2015, methane emissions have fallen significantly. From the perspective of evolution, the No.1 central document in 2004 "opinions on several policies of promoting farmers' income" was put forward after 18 years, and its main body returned to "the field of agriculture, rural areas and farmers". The document put forward the "three subsidies" of direct subsidies for farmers, subsidies for improved varieties, and subsidies for agricultural machinery. It also proposed some measures to reduce the agricultural tax burden, which effectively enhanced the enthusiasm of farmers to grow grain and prompted significant improvement in the level of green development in 2006. In 2015, in order to cope with the problems such as the "ceiling" of prices, the rise of the "floor" of production costs, and the intensified challenges of the "hard constraints" of resources and the environment, the General Office of the State Council of the People's Republic of China issued "the Opinions on Accelerating the Transformation of Agricultural Development Mode", which provided an important starting point for promoting agricultural development on quantity, quality, and efficiency. Under the guidance of the document, the structural adjustment to the agriculture led to a further decline in methane emissions. In terms of the sources of methane emissions, from 2000 to 2019, emissions from paddy fields fell from 6.51 million tons to 6.29 million tons, emissions from livestock and poultry manure decreased from 2.72 million tons to 1.88 million tons, and emissions from livestock and poultry intestinal fermentation dropped from 12.23 million tons to 9.53 million tons (Figure 2). The decline of livestock and poultry intestinal emissions played a crucial role in reducing agricultural methane emissions.

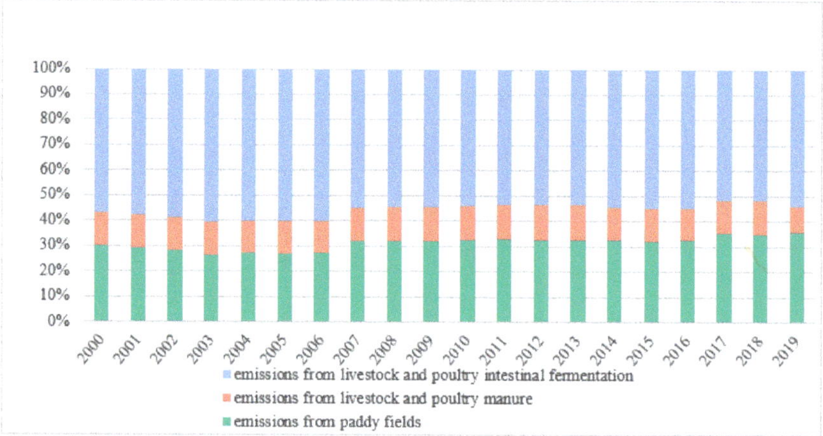

Figure 2. Agricultural methane emissions percentage in China, 2000–2019.

3.1.2. Spatial Analysis of Different Production Areas

Considering the differences between agricultural development among different regions in China, the analysis between spatial regions is more targeted. Based on the classification of agricultural grain functional areas, the whole country is divided into main producing area, main sales area, and balance area.

Agricultural methane emissions in the main producing area, main sales area, and balance area showed a downward trend, but the changes had their own characteristics (Figure 3). The main producing area was the region with the highest agricultural methane emissions. From 2000 to 2019, the main producing area decreased the most, from 12.79 million tons to 10.84 million tons. From the perspective of the planting industry, the sown area of double-cropping early rice and double-cropping late rice in the main producing areas decreased by 27.61% and 31.83%, respectively, but the sown area of single-crop rice increased by 48.38%, which led to an increase in methane emissions from paddy fields in the main producing areas. From the perspective of the breeding industry, although the methane emissions from paddy fields in the main producing areas increased to a certain extent, an increase of 14.36%, the methane emissions from livestock and poultry manure management and enteric fermentation decreased by 507,500 tons and 2,028,000 tons, respectively. From 2000 to 2019, the number of livestock and poultry breeding dropped significantly, and the number of non-dairy cows, goats, pigs, poultry, horses, camels, and donkeys decreased by 43.83%, 20.43%, 29.04%, 16.78%, 62.97%, 88.65%, and 70.96%, respectively. The number of poultry breeding fell the most. The main production areas bear the heavy responsibility of China's food security, and the decline in emissions shows that the industrial structure has been improved, to a certain extent, and agricultural green production has gradually improved. From the perspective of planting industry, the sown area of double-cropping early rice and double-cropping late rice in the main producing areas decreased by 27.61% and 31.83%, respectively, but the sown area of single-crop rice increased by 48.38%, which led to an increase in methane emissions from paddy fields in the main producing areas. From the perspective of the breeding industry, although the methane emissions from paddy fields in the main producing areas increased to a certain extent, with an increase of 14.36%, the methane emissions from livestock and poultry manure management and enteric fermentation decreased by 507,500 tons and 2,028,000 tons, respectively. From 2000 to 2019, the number of livestock and poultry breeding dropped significantly, and the number of non-dairy cows, goats, pigs, poultry, horses, camels, and donkeys decreased by 43.83%, 20.43%, 29.04%, 16.78%, 62.97%, 88.65%, and 70.96%, respectively, and the number of poultry breeding fell the most. The main production areas bear the heavy responsibility of China's food security, and the decline in emissions has shown the improved industrial structure and increased agricultural green production. Methane emissions in the balance area fell from 6.29 million tons to 5.60 million tons. In the balance area, methane emissions from rice fields, livestock manure management, and livestock intestinal fermentation all decreased by 23,600 tons, 169,800 tons and 285,400 tons, respectively. The balance area is located mainly in Central and Western China, and the change in emissions was partly due to the implementation of China's western development strategy. In addition, methane emissions in the main sales area dropped from 2.37 million tons to 1.25 million tons, decreasing by 47.48%. The methane emissions from paddy fields, livestock manure management, and livestock enteric fermentation in the main sales area decreased by 41%, 48.14%, and 61.25%, respectively. Except for dairy cows, the number of livestock and poultry has decreased significantly. This was mainly due to the fact that the main sales areas, such as Beijing, Tianjin, and Shanghai, are relatively developed areas in China. With limited land and other planting resources and a large food shortage, it is necessary for these developed areas to rely on other provinces to transport food, forcing the region to adjust the industrial structure and objectively promoting the reduction of agricultural methane emissions. The decline in methane emissions in main producing and sales regions was the main reason for the overall decline. Therefore, the main producing areas have high methane emissions and are key areas for the green transformation and development of agriculture.

From 2000 to 2019, the emissions of the main producing areas, main sales areas, and balanced areas were all trending downward, of which the largest decline rate was the main sales area (47.48%), and the largest decline amount was the main producing area (1.9539 million tons). Although they all showed a downward trend, the changes had their own characteristics. For example, the main grain producing areas were the areas with the highest agricultural methane emissions which have dropped by approximately 15.27% in the past 20 years, with an average annual decline of 0.83%, of which 2000–2005 and 2010–2015 were the rising periods, with increases of 10.78% and 2.34%, respectively; 2005–2010 and 2015–2019 were the decreasing periods, and the decline was relatively large, with 13.14% and 13.96%, respectively. Although the total amount of emissions had a downward trend, the discharge of paddy fields had an upward trend, and the proportion of emissions was also different from the total amount of emissions. 2000–2005 and 2015–2019 were decreasing periods, 2005–2015 was the rising period, where the total amount increased from 4.0507 million tons in 2000 to 4.6324 million tons in 2019, and the proportion of emissions increased from 31.66% to 42.73%. Manure management emissions and intestinal fermentation emissions were on a downward trend, decreasing by approximately 30.02% and 28.75%, respectively. In short, the emission structure of the main producing areas was optimized, where livestock and poultry intestinal emissions were the main source of emissions; the proportion of paddy field emissions and intestinal emissions was close, with 42.73% and 46.36%, respectively in 2019. Paddy field management will continue to be an important measure to reduce emissions in major producing areas. Agricultural methane emissions from the main sales areas were the lowest, and although the agricultural methane emissions from paddy fields were not the highest among the three grain divisions, the ratio of emission sources in the main sales areas was the highest, from 58.67% in 2000 to 65.91% in 2019. In addition, 2000–2005 was a decline period, reducing from 58.67% to 52.99%, followed by an increase period in 2005–2019, with a growth of 8.91%. The trend of changes in the proportion of manure management emissions and intestinal emissions was similar, showing the trend of first rising and then declining, of which the total amount of manure management emissions did not change much in 2000–2010, but the proportion of emissions increased from 14.42% to 19.32%. Intestinal fermentation emissions fell by 23,500 tons between 2000 and 2005, but the proportion of emissions increased from 26.91% to 29.99%, and then in 2019, the proportion of emissions fell to 19.85%. Overall, intestinal fermentation emissions fell by 391,300 tons, with the proportion decreasing by 7.05%. The total decline (691,200 tons) and the decline rate (10.99%) in the balance area in 20 years were the lowest, similar to the change trajectory of the main producing areas where the increase periods were 2000–2005 and 2010–2015, and the decline periods were 2005–2010 and 2015–2019. The emission sources were similar to those in the main producing areas, with intestinal fermentation emissions as the main emission source, but the change amplitude of the three emission sources was not high. In addition, it should be noted that the proportions of intestinal emissions and paddy field emissions were different from the change trend of the main producing areas and the main sales areas; the change range from 2000 to 2019 was between 3.81% and −2.12%, and the three emission sources were reduced by 236,000 tons, 169,800 tons, and 285,400 tons, respectively, from which can be seen that the emission reduction effect of the balance area was not good. In addition, in 2000, the main producing area and the main sales area were 5.39 times and 2.65 times, respectively, the total emissions of the balanced area and by 2019, they had developed into 8.69 times and 4.49 times, respectively; this, to a certain extent, also reflects the widening of spatial differences.

3.1.3. Spatial Analysis of Different Provinces

In view of the different emphases of industrial structure and economic development in each province, the changes in agricultural methane emissions in each province are analyzed at the provincial level (Figure 4). In 2000, the top provinces in agricultural methane emissions were Hunan, Henan, Sichuan, Guangxi, and Shandong, and by 2010, the top provinces were Hunan, Sichuan, Henan, Inner Mongolia, and Guangxi. The emissions of Shandong decreased from 5.70% to 3.90%, and the emissions of Inner Mongolia increased from 3.27% to 5.68%. In addition, Heilongjiang rose significantly from 3.19% to 5.01%. In 2019, the top emission provinces were Hunan, Sichuan, Inner Mongolia, Heilongjiang, and Jiangxi. The reason for the highest agricultural methane emissions in Hunan was that it is a large paddy production and marketing province in China, with the paddy planting area of 3.8556 million hm2 in 2019. The large-scale breeding rate of cattle and sheep in Sichuan Province was low, and the proportion of large-scale pig breeding in the first half of 2020 (53.8%) exceeded the free-range ratio (46.2%) for the first time, resulting in high intestinal methane emissions from livestock and poultry; however, intensive production is gradually becoming the mainstream of production, and greening has been further improved. In 2018, Henan guided farmers to implement the plan of advocating intensive agricultural production, and at the end of 2018, 980,000 livestock and poultry free-range households had been "withdrawn from the village", which caused agricultural methane emissions in Henan to decrease by 2.31% from 2010 to 2019. The number of sheep raised in Inner Mongolia was much higher than that of other provinces; the large-scale breeding rate needs to be improved, which was the main reason for the high emission of agricultural methane. Therefore, scale is the only way for the transformation and development of animal husbandry. It is impossible for retail farming to completely withdraw, and financial means should be strengthened to support and promote new industrial models. For example, building a comprehensive platform for investment funds and technologies by industry experts and attracting retail investors to join the company can not only reduce the risk and cost of retail investors but also increase the rate of large-scale breeding and effectively reduce methane emissions. The agricultural methane emissions of Tianjin, Beijing, Shanghai, and other Eastern developed provinces were relatively small and remained at the bottom of the emission proportion in 2000, 2010, and 2019. The reason lay in the composition of industrial structure. For example, in 2019, the proportion of the first industry in the GDP of the three provinces was 1.3%, 0.3%, and 0.3%, respectively. The number of livestock and poultry in Ningxia and Hainan was low; in 2019, the number of livestock and poultry was 20.707 million and 63.913 million, respectively, far lower than the national average of 233.2885 million, so the methane emission was low. Liaoning and Jilin, also major grain-producing regions in Northeast China, had relatively low emissions, ranking lower (20th and 21st, respectively, in 2019). Compared with Heilongjiang, the two provinces had a lower paddy sown area, with only 22% and 13%, respectively, than Heilongjiang in 2019. In addition, among the main producing areas, Hebei accounted for 4.05% of emissions in 2000, and the number of livestock and poultry decreased by 37.26% in 2010 compared with 2000, which was the main reason for its low emissions, ranking 18th in 2019.

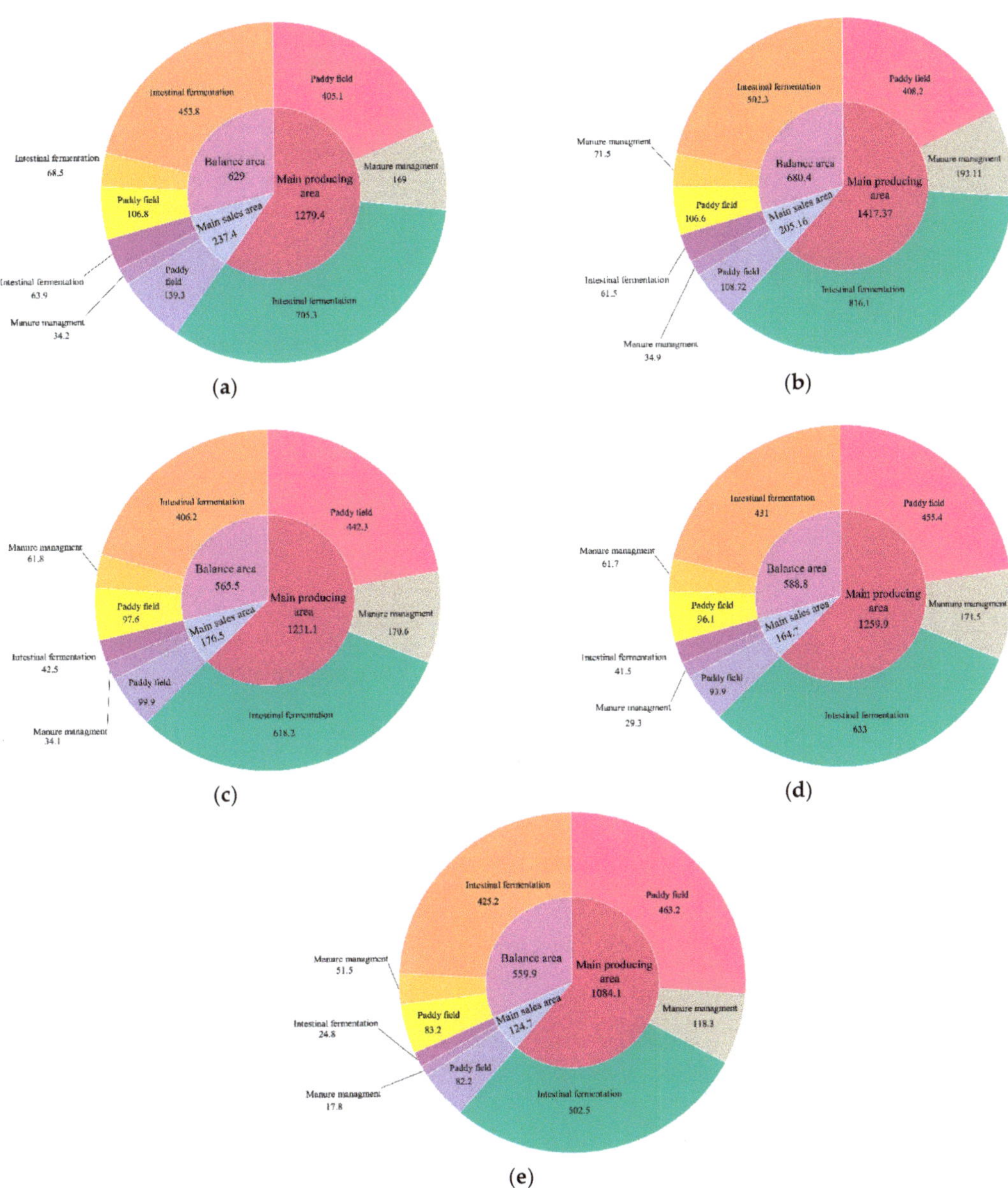

Figure 3. Agricultural methane emissions by region in China, 2000–2019. (**a**) 2000; (**b**) 2005; (**c**) 2010; (**d**) 2015; and (**e**) 2019.

From the perspective of emission sources, in 2019, the agricultural methane emission in 12 of the 31 provinces was dominated by paddy field emissions, represented by Jiangsu (82.73%), and the remaining 19 provinces were dominated by intestinal fermentation emissions, represented by Qinghai, Gansu, Tibet, Xinjiang, and Inner Mongolia, which accounted for more than 90%, and the intestinal fermentation emissions caused by them reached 18.57% of the national total emissions.

3.1.4. Dynamic Evolution of Spatial Differentiation

To effectively reflect the spatial characteristics of agricultural methane emissions, the dynamic distribution of 2000–2019 and 2010–2019 was analyzed by non-parametric kernel density. As can be seen from Figure 5, the kernel density showed two peaks in the two time periods, and then decreased continuously. The distribution curve showed the right tail, the peak moved significantly to the right, and the gap of provincial agricultural methane emissions showed a widening trend.

3.2. Influencing Factors of Agricultural Methane Emission in China

3.2.1. Model Building of Influencing Factors

Referring to relevant references, the factors affecting agricultural methane emissions include agricultural development level, residents' living conditions, industrial structure, urbanization level, industrialization level, environmental regulation, agricultural machinery input, and agricultural science and technology R&D investment, etc.

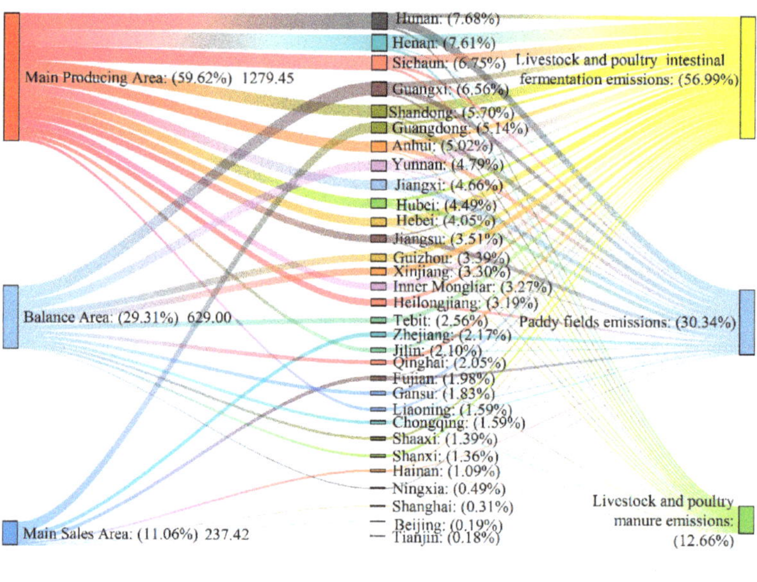

(a)

Figure 4. *Cont.*

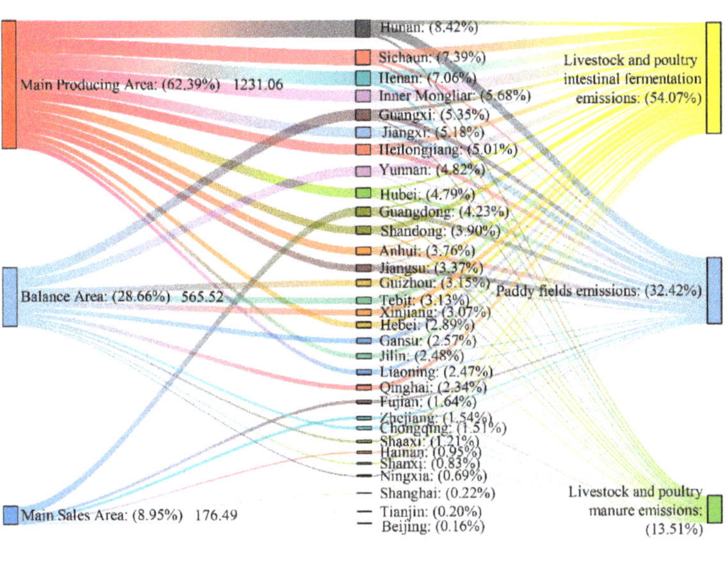

(**b**)

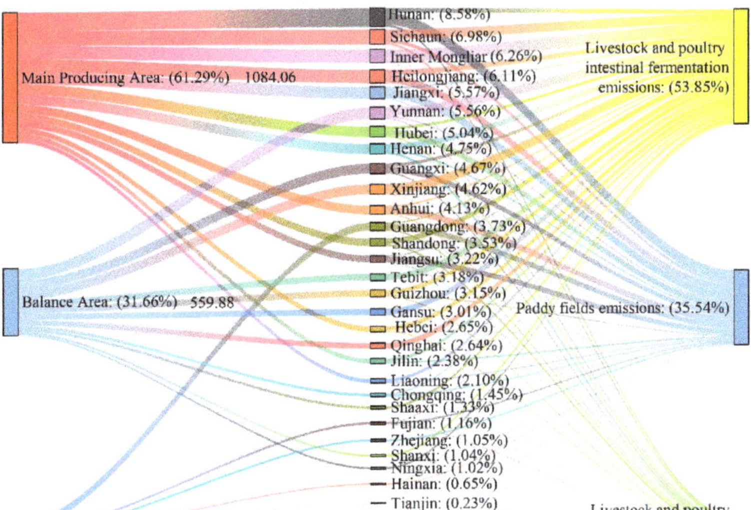

(**c**)

Figure 4. Agricultural methane emissions by provinces in China, 2000–2019. (**a**) 2000; (**b**) 2010; and (**c**) 2019.

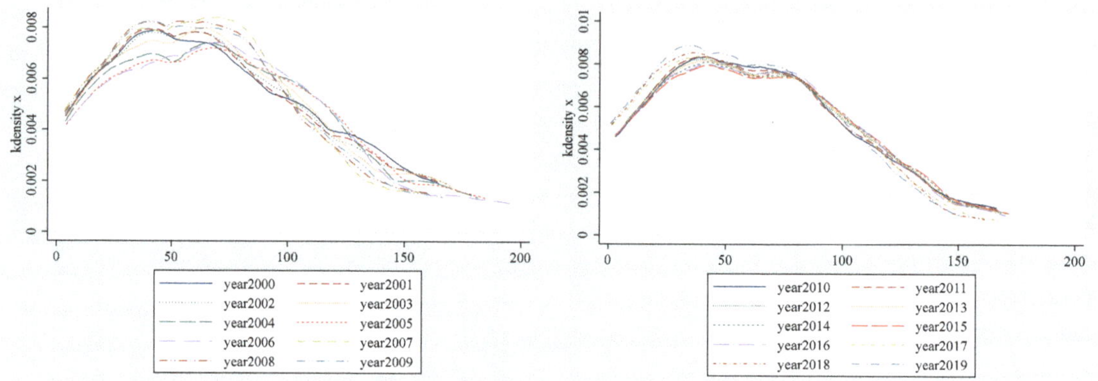

Figure 5. Epanechnikov kernel density function diagram.

Agricultural development level. There is a double-sided effect between agricultural development level and agricultural methane emissions. The improvement of agricultural development level is often accompanied by the expansion of production scale or structural adjustment, which may produce more methane emissions in this process. On the contrary, the improvement of agricultural development level drives the improvement of infrastructure construction and agricultural production capacity and promotes the decline of methane emissions. In this paper, (agr) is used to represent the per capita gross output value of agriculture, forestry, animal husbandry and fishery. Residents' living conditions. There is an "inverted U-shaped" relationship between residents' living conditions and agricultural methane emissions. With the improvement of residents' living conditions, a greener and better ecological environment is needed. Therefore, there is a negative relationship between them, which is represented by (inc), the per capita income of rural residents. Industrial structure. Methane emissions come from planting and breeding. Therefore, the internal structure of the agricultural industry will affect methane emissions, which is represented by (ind), the proportion of planting industry in the total output value of agriculture, forestry, animal husbandry, and fishery. Urbanization level. Urbanization will occupy agricultural land, thus reducing agricultural methane emissions, which is measured by the proportion of urban population in total population and represented by (ubr). Industrialization level. With the development of industrialization, the emergence of green technology will reduce methane emissions in the process of agricultural production, which is represented by (iind), the proportion of non-agricultural total output value. Environmental regulations. Environmental regulations will have an impact on agricultural production and will encourage farmers engaged in agricultural production, especially aquaculture production, to adopt green technology and improve new technologies in the process of agricultural production. In this paper, (enr) is used to characterize the proportion of pollution control projects completed in each region in the year in the regional GDP. Agricultural machinery input. The input of agricultural machinery is mainly used to reduce methane emissions in agricultural planting. The input of great horsepower machinery can effectively improve planting efficiency, which is represented in this paper by (am), the total power of agricultural machinery. Agricultural science and technology R&D investment. Investment in agricultural science and technology R&D is the key to promoting agricultural technology progress. In this paper, (ati) is used to represent the proportion of government investment and financial support to agriculture in the total expenditure of provincial researchers.

3.2.2. Analysis of Influencing Factors of Spatial Differences of Agricultural Methane

On the basis of the geographical detector, this paper takes the three grain production areas as the main hierarchical method, uses the natural breakpoint method to treat variables by stratification, and divides the impact factors into four categories(Table 4). It can be seen that 9 factors play a driving role in the spatial differentiation of methane emissions, and that the influencing factors are divided into key and secondary influence factors according to the size of the driving factor explanatory power (q-statistical value). Among them, spatial factors (spa), industrial structure status (ind), industrialization level (iind), and agricultural science and technology R&D investment (ati) are the key influencing factors; agricultural development level (agr), residents' living conditions (inc), urbanization level (ubr), environmental regulation (enr), and agricultural machinery input (am) are the secondary influencing factors. The explanatory power of spatial factors reaches 0.438, which is the most key influencing factor, indicating that the production differences in different grain production zones played a decisive role in agricultural methane emissions. The explanatory power of industrial structure, industrialization level, and agricultural R&D investment are 0.122, 0.241, and 0.206, respectively. These variables represent the driving effect of agricultural transformation and upgrading on agricultural methane reduction from multiple aspects. In addition, the explanatory power of residents' living conditions, urbanization level, environmental regulation, and agricultural machinery input are all above 0.03, which cannot be ignored even though it is not the most critical factor influencing regional differences during the investigation period.

Table 4. Detection results of influencing factors of agricultural methane emissions in China.

Factor	spa	agr	inc	ind	ubr	iind	enr	am	ati
q	0.438	0.041	0.044	0.122	0.034	0.241	0.032	0.082	0.206
p	0.000	0.005	0.017	0.000	0.038	0.000	0.035	0.000	0.000

3.2.3. Influencing Factors of Agricultural Methane Emission in Different Production Areas

Agricultural science and technology R&D investment has an impact on the spatial distribution of the three regions. It provides important technical support for methane emission reduction, and continuously strengthening the government's investment in agricultural science and technology is an important driver to achieve green development. The key influencing factors of main producing area, main sales area, and balance area are different(Table 5). Specifically, for the main producing area, the industrial structure is the most critical factor, and its explanatory power can reach 0.549. Main producing areas undertake the due obligations to ensure food security. Under the safety bottom line, adjusting the proportion between agriculture, forestry, animal husbandry, and fishery, especially the reasonable industrial structure adjustment process under the guidance of nutrition, will play an important role in methane emission reduction. The explanatory power of industrialization level and agricultural machinery input is also above 0.08, which is the main influencing factor of the spatial distribution of agricultural methane emission in main producing areas.

Table 5. Factor detection results of methane emission in different production areas.

Factor	Main Producing Area		Main Sales Area		Balance Area	
	q	p	q	p	q	p
agr	0.010	0.694	0.292	0.002	0.132	0.045
inc	0.021	0.509	0.102	0.038	0.025	0.478
ind	0.549	0.000	0.055	0.141	0.479	0.000
ubr	0.009	0.816	0.029	0.403	0.047	0.196
iind	0.083	0.056	0.038	0.289	0.545	0.000
enr	0.049	0.107	0.083	0.127	0.042	0.234
am	0.083	0.042	0.178	0.029	0.036	0.319
ati	0.071	0.031	0.106	0.085	0.195	0.000

For the main sales area, agricultural development level is the most important factor, and its explanatory power reaches 0.292. The level of agricultural development in main sales area will affect the income of residents and their demand for a better life. The better the agricultural development level, the stronger the demand for green and low-carbon products, which will reduce the methane emissions. The explanatory power of residents' living standards and agricultural mechanization input is also higher than 0.1, indicating that these factors are also important factors influencing the spatial distribution of agricultural methane.

For the balance area, the industrial structure is the main influencing factor with the explanatory power of 0.479, and the industrialization level is also the main influencing factor with the explanatory power of 0.545. The evolution of industrial structure is that the transformation from planting to the integration of primary, secondary, and tertiary industries, which will effectively reduce agricultural methane emissions. The acceleration of industrialization will shift the rural labor force and release a greater demand to replace chemical fertilizers with organic fertilizer. Agricultural development level and investment in agricultural science and technology R&D are also key factors affecting the spatial distribution of balance area.

4. Discussion

Agriculture is both a source of greenhouse gas emissions and one of the most sensitive sectors to the impact of temperature change. In recent years, agricultural methane emission reduction has achieved some success, but the impact of greenhouse gas emissions caused by agricultural activities on climate change cannot be ignored. The year 2019 was the fifth warmest year since 1951, the national average temperature was 0.79 °C higher than usual, the average number of high temperature days was 11.8 days, with 4.1 days more than usual, compared with the second warmest year. Methane is the second largest greenhouse gas, and the recyclability of methane shows that controlling agricultural methane emission is an effective way to achieve the dual carbon goal. From the mechanism of agricultural methane emission, the formation ways of agricultural methane include paddy field emissions from the planting industry, livestock and poultry manure emission from the breeding industry, and livestock and poultry intestinal fermentation emission. Among them, the emission of livestock and poultry intestinal fermentation is the main emission source, accounting for 51.55–60.71%. Therefore, it is important to focus on controlling the emission of livestock and poultry intestinal fermentation to control agricultural methane emission, adjust the livestock breeding structure according to the actual situation of each region, and promote standardized breeding is an effective means. Paddy field emissions from planting industry is the second largest emission source after livestock and poultry intestinal fermentation, accounting for 26.45–35.54%. As the largest paddy producing country, China should choose appropriate paddy varieties and cultivation methods, promote large-scale planting, improve the utilization capacity of straw returning to the field, and other measures to reduce the emission coefficient of paddies. Compared with the previous literature, this paper adds the analysis of grain functional areas, uses nuclear density to investigate the differences and changes, utilizes geographic detectors to analyze the influencing factors, and then puts forward targeted emission reduction measures and countermeasures.

On the basis of the above analysis, the countermeasures to reduce agriculture methane emissions are provided as follows.

Firstly, it is important to carry out spatial cooperation and explore the spatiotemporal absorptive mechanism of agricultural methane emission reduction. The administrative boundary should be broken, the spatial association network should be constructed, and the free flow of elements (manpower, capital, etc.) in space should be realized. In addition, the temporal and spatial benefit protection and incentive policy support system needs to be constructed. Initially, it is necessary to clarify the spatial constraint characteristics of agricultural development, especially the differences between natural resources endowments. From the national level, legal, economic, and social tools should be adopted to provide institutional guarantee for the coordination mechanism of cross-regional green development of agriculture. Then, a national cross-regional cooperation platform should be established. Consistent with the development of the national carbon market, the platform of methane emission reduction also needs to be built. In terms of agricultural methane emission reduction, attention should be paid to reducing the differences between regions, such as setting up a regional agricultural methane emission reduction fund, to form a platform for cooperation and exchange between regions.

Secondly, it is crucial to strengthen the research and development of green and low-carbon agricultural technologies. The investment in agricultural science and technology has a significant impact on the spatial distribution of agricultural methane in the three grain functional areas; therefore, increasing investment in agricultural science and technology research and development is the most effective measure to form China's unique development pattern of high-value agriculture and low-carbon agriculture. In the input, the role of the promising government and the effective market shall be highlighted, social capital shall be guided into basic research and development, and an integrated guidance mechanism of breeding, reproduction, and promotion shall be established. In the output, the enthusiasm of various agricultural business entities needs to be stimulated, and a new pattern of green development should be formed by combining key technologies such as big data and smart agriculture. In addition, methane can be further utilized by recycling. The liquid storage process can be reduced for manure with high emission potential, and methane can be recovered through anaerobic fermentation to reduce greenhouse gas emissions. Methane can be recycled through the construction of biogas projects, and emissions can be reduced by changing the wet manure into dry manure and changing the way of manure is stored by covering.

Thirdly, the mechanism of reducing agricultural methane emission by time, land, and classification should be formed. For main producing area and balance area, emphasis should be placed on the optimization of industrial structure, especially on the premise of ensuring food security to promote the green transformation of industrial structure. The adjustment of industrial structure should be carried out from the aspects of spatial distribution, product structure, and industrial chain upgrading. The spatial layout should mainly highlight local characteristics, optimize the spatial layout, take into account variety improvement and quality improvement in product structure, improve regional quality agricultural products, solve the problem of "high quality but not good price", upgrade the industrial chain to achieve vertical extension, and promote advanced planting and breeding, advanced planting and fishing, ecological circular agriculture development, and realize the integrated development of primary, secondary, and tertiary industries. For the main grain sales area, the leading function of agricultural green transformation should be brought into play, and residents' awareness of green agricultural products should be initiated to drive the improvement and promotion of the entire agricultural industry chain.

This paper selects the emission coefficient published by IPCC for calculation, which is a widely used research method; however, due to the different actual conditions in various regions of the country, there is a problem of low accuracy in the face of the change of emission coefficient. At the same time, this paper lacks specific analysis at the provincial level, which is the direction of improvement in the future.

5. Conclusions

On the basis of the construction of the agricultural methane measurement method, the total amount and structure of the agricultural methane emissions in China is calculated and the spatiotemporal differentiation features of the agricultural methane emission through the grain production areas are analyzed. Then, through the interpretation of the kernel density function curve, distribution location, sample state, and kurtosis extension—the key factors affecting the spatial differentiation of agricultural methane emission—are analyzed by using geographic detectors. The main conclusions are as follows.

Firstly, agricultural methane emissions showed a trend of increasing first and then decreasing from 2000 to 2019; agricultural methane emissions showed a downward trend in 2006 and have been stable since then. During the 13th Five-Year Plan, the greening level was further improved, and methane emissions declined steadily. Agricultural methane emission reduction is closely related to the green development policy.

Secondly, the characteristics of agricultural methane emissions in the three major grain areas are basically the same as those in the whole country, but the changes in the main producing area, main sales area, and balance area have their own characteristics. From 2000 to 2019, the main reason for the reduction of agricultural methane emissions in the main producing area was the decline of intestinal fermentation emissions of livestock and poultry, which exceeded the increase of emissions from paddy fields. The paddy field, livestock and poultry intestinal fermentation, and manure management emissions in the main sales area and balance area fell, which generally reflects that the overall effect of green development in China is gradually emerging. At the provincial level, emphasis should be placed on reducing agricultural methane emissions in major producing provinces such as Hunan, Sichuan, Inner Mongolia, Heilongjiang, and Jiangxi.

Thirdly, the spatial patterns of agricultural methane emissions in China will remain unchanged in the short term, and it is difficult to achieve convergence.

Fourthly, the production heterogeneity of grain production areas has a decisive impact on the difference in agricultural methane emissions. Industrial structure, the level of industrialization, the investment in agricultural science, and technology R&D are the key influencing factors. Agricultural development level, residents' living conditions, urbanization level, environmental regulation, and agricultural machinery input are secondary influencing factors. In particular, the investment in agricultural science and technology R&D is influential in the agricultural methane emission pattern in the three main grain areas. For the main grain sales area, the agricultural development level is the core driving factor for the main sales area, and the industrialization level and industrial structure are the main driving factors for the balance area.

Author Contributions: Conceptualization, J.H.; data curation, P.L.; formal analysis, P.L.; investigation, J.H.; methodology, G.W.; writing—original draft, G.W.; writing—review & editing, F.Z. All authors have read and agreed to the published version of the manuscript.

Funding: This research was funded by the Science Fund for Creative Research Groups of the National Natural Science Foundation of China (Grant No. 72221002) and the National Natural Science Foundation of China [72003111].

Institutional Review Board Statement: Not applicable.

Data Availability Statement: Data available from the authors upon request.

Conflicts of Interest: The authors declare no conflict of interest.

References

1. Froese, R.; Schilling, J. The nexus of climate change, land use, and conflicts. *Curr. Clim. Chang. Rep.* **2019**, *5*, 24–35. [CrossRef]
2. Godde, C.M.; Boone, R.B.; Ash, A.J.; Waha, K.; Sloat, L.L.; Thornton, P.K.; Herrero, M. Global rangeland production systems and livelihoods at threat under climate change and variability. *Environ. Res. Lett.* **2020**, *15*, 044021. [CrossRef]
3. Hrabok, M.; Delorme, A.; Agyapong, V.I. Threats to mental health and well-being associated with climate change. *J. Anxiety Disord.* **2020**, *76*, 102295. [CrossRef] [PubMed]
4. Nhemachena, C.; Nhamo, L.; Matchaya, G.; Nhemachena, C.R.; Muchara, B.; Karuaihe, S.T.; Mpandeli, S. Climate change impacts on water and agriculture sectors in Southern Africa: Threats and opportunities for sustainable development. *Water* **2020**, *12*, 2673. [CrossRef]
5. Mitchell, D.; Allen, M.R.; Hall, J.W.; Muller, B.; Rajamani, L.; Le Quéré, C. The myriad challenges of the Paris Agreement. *Philos. Trans. R. Soc. A Math. Phys. Eng. Sci.* **2018**, *376*, 20180066. [CrossRef]
6. Winkler, H.; Mantlana, B.; Letete, T. Transparency of action and support in the Paris Agreement. *Clim. Policy* **2017**, *17*, 853–872. [CrossRef]
7. Li, H.-M.; Wang, X.-C.; Zhao, X.-F.; Qi, Y. Understanding systemic risk induced by climate change. *Adv. Clim. Chang. Res.* **2021**, *12*, 384–394.

8.	Li, Y.; Wei, Y.; Zhang, X.; Tao, Y. Regional and provincial CO_2 emission reduction task decomposition of China's 2030 carbon emission peak based on the efficiency, equity and synthesizing principles. *Struct. Chang. Econ. Dyn.* **2020**, *53*, 237–256. [CrossRef]

9.	Chen, X.; Shuai, C.; Wu, Y.; Zhang, Y. Analysis on the carbon emission peaks of China's industrial, building, transport, and agricultural sectors. *Sci. Total Environ.* **2020**, *709*, 135768. [CrossRef]

10.	Dean, J.F. Old methane and modern climate change. *Science* **2020**, *367*, 846–848. [CrossRef]

11.	Kholod, N.; Evans, M.; Pilcher, R.C.; Roshchanka, V.; Ruiz, F.; Coté, M.; Collings, R. Global methane emissions from coal mining to continue growing even with declining coal production. *J. Clean. Prod.* **2020**, *256*, 120489. [CrossRef] [PubMed]

12.	IPCC. Climate Change 2021: The Physical Science Basis. In *Contribution of Working Group I to the Sixth Assessment Report of the Intergovernmental Panel on Climate Change*; Cambridge University Press: Cambridge, UK; New York, NY, USA, 2021.

13.	International Energy Agency. World Energy Outlook 2017. Available online: https://www.iea.org/weo2017/ (accessed on 14 November 2017).

14.	Derwent, R.G. Global warming potential (GWP) for methane: Monte Carlo analysis of the uncertainties in global tropospheric model predictions. *Atmosphere* **2020**, *11*, 486. [CrossRef]

15.	Ocko, I.B.; Sun, T.; Shindell, D.; Oppenheimer, M.; Hristov, A.N.; Pacala, S.W.; Mauzerall, D.L.; Xu, Y.; Hamburg, S.P. Acting rapidly to deploy readily available methane mitigation measures by sector can immediately slow global warming. *Environ. Res. Lett.* **2021**, *16*, 054042. [CrossRef]

16.	Skytt, T.; Nielsen, S.N.; Jonsson, B.-G. Global warming potential and absolute global temperature change potential from carbon dioxide and methane fluxes as indicators of regional sustainability—A case study of Jämtland, Sweden. *Ecol. Indic.* **2020**, *110*, 105831. [CrossRef]

17.	Rehman, A.; Ma, H.; Irfan, M.; Ahmad, M. Does carbon dioxide, methane, nitrous oxide, and GHG emissions influence the agriculture? Evidence from China. *Environ. Sci. Pollut. Res.* **2020**, *27*, 28768–28779. [CrossRef]

18.	Smith, P.; Reay, D.; Smith, J. Agricultural methane emissions and the potential formitigation. *Philos. Trans. R. Soc. A* **2021**, *379*, 20200451. [CrossRef]

19.	Seketeme, M.; Madibela, O.R.; Khumoetsile, T.; Rugoho, I. Ruminant contribution to enteric methane emissions and possible mitigation strategies in the Southern Africa Development Community region. *Mitig. Adapt. Strateg. Glob. Chang.* **2022**, *27*, 1–26. [CrossRef]

20.	Twinkle Devi, N.; Sarjubala Devi, A.; Singh, K.R. Emission of Methane From Wetland Paddy Fields: A Review. *J. Clim. Chang.* **2022**, *8*, 13–19. [CrossRef]

21.	Kuhla, B.; Viereck, G. Enteric methane emission factors, total emissions and intensities from Germany's livestock in the late 19th century: A comparison with the today's emission rates and intensities. *Sci. Total Environ.* **2022**, *848*, 157754. [CrossRef]

22.	Haider, A.; Bashir, A.; ul Husnain, M.I. Impact of agricultural land use and economic growth on nitrous oxide emissions: Evidence from developed and developing countries. *Sci. Total Environ.* **2020**, *741*, 140421. [CrossRef]

23.	Hou, P.; Yu, Y.; Xue, L.; Petropoulos, E.; He, S.; Zhang, Y.; Pandey, A.; Xue, L.; Yang, L.; Chen, D. Effect of long term fertilization management strategies on methane emissions and rice yield. *Sci. Total Environ.* **2020**, *725*, 138261. [CrossRef] [PubMed]

24.	Liu, Y.; Tang, H.; Muhammad, A.; Huang, G. Emission mechanism and reduction countermeasures of agricultural greenhouse gases–a review. *Greenh. Gases Sci. Technol.* **2019**, *9*, 160–174. [CrossRef]

25.	Lynch, J.; Garnett, T. Policy to reduce greenhouse gas emissions: Is agricultural methane a special case? *EuroChoices* **2021**, *20*, 11–17. [CrossRef]

26.	Tarazkar, M.H.; Kargar Dehbidi, N.; Ansari, R.A.; Pourghasemi, H.R. Factors affecting methane emissions in OPEC member countries: Does the agricultural production matter? *Environ. Dev. Sustain.* **2021**, *23*, 6734–6748. [CrossRef]

27.	Wang, Z.; Zhang, X.; Liu, L.; Wang, S.; Zhao, L.; Wu, X.; Zhang, W.; Huang, X. Estimates of methane emissions from Chinese rice fields using the DNDC model. *Agric. For. Meteorol.* **2021**, *303*, 108368. [CrossRef]

28.	Zhang, X.Z.; Wang, J.Y.; Zhang, T.L.; Li, B.L.; Yan, L. Assessment of methane emissions from China's agricultural system and low carbon measures. *Environ. Sci. Technol.* **2021**, *44*, 200–208.

29.	Li, Y.; Chen, M.P. Influencing factors of methane and nitrous oxide emissions from agricultural sources in China. *Acta Sci. Circumstantiae* **2021**, *41*, 710–717.

30.	Pu, Y.; Zhang, M.; Jia, L.; Zhang, Z.; Xiao, W.; Liu, S.; Zhao, J.; Xie, Y.; Lee, X. Methane emission of a lake aquaculture farm and its response to ecological restoration. *Agric. Ecosyst. Environ.* **2022**, *330*, 107883. [CrossRef]

Article

Soil Autotrophic Bacterial Community Structure and Carbon Utilization Are Regulated by Soil Disturbance—The Case of a 19-Year Field Study

Chang Liu [1,2], Junhong Xie [1,2], Zhuzhu Luo [1,3], Liqun Cai [1,3] and Lingling Li [1,2,*]

1   State Key Laboratory of Aridland Crop Science, Gansu Agricultural University, Lanzhou 730070, China
2   College of Agronomy, Gansu Agricultural University, Lanzhou 730070, China
3   College of Resource and Environment, Gansu Agricultural University, Lanzhou 730070, China
*   Correspondence: lill@gsau.edu.cn

Citation: Liu, C.; Xie, J.; Luo, Z.; Cai, L.; Li, L. Soil Autotrophic Bacterial Community Structure and Carbon Utilization Are Regulated by Soil Disturbance—The Case of a 19-Year Field Study. *Agriculture* **2022**, *12*, 1415. https://doi.org/10.3390/agriculture12091415

Academic Editor: Luciano Beneduce

Received: 21 July 2022
Accepted: 20 August 2022
Published: 8 September 2022

Abstract: The roles of bacterial communities in the health of soil microenvironments can be more adequately defined through longer-term soil management options. Carbon dioxide (CO_2) fixation by autotrophic bacteria is a principal factor in soil carbon cycles. However, the information is limited to how conservation tillage practices alter soil physiochemical properties, autotrophic bacterial communities, and microbial catabolic diversity. In this study, we determined the changes in autotrophic bacterial communities and carbon substrate utilization in response to different soil management practices. A replicated field study was established in 2001, with the following soil treatments arranged in a randomized complete block: conventional tillage with crop residue removed (T), conventional tillage with residue incorporated into the soil (TS), no tillage with crop residue removed (NT), and no tillage with residue remaining on the soil surface (NTS). Soils were sampled in 2019 and microbial DNA was analyzed using high-throughput sequencing. After the 19-year (2001–2019) treatments, the soils with conservation tillage (NTS and NT) increased the soil's microbial biomass carbon by 13%, organic carbon by 5%, and total nitrogen by 16% compared to conventional tillage (T and TS). The NTS treatment increased the abundance of the *cbbL* gene by 53% in the soil compared with the other soil treatments. The *cbbL*-carrying bacterial community was mainly affiliated with the phylum *Proteobacteria* and *Actinobacteria*, accounting for 56–85% of the community. Retaining crop residue in the field (NTS and TS) enhanced community-level physiological profiles by 31% and carbon substrate utilization by 32% compared to those without residue retention (T and NT). The 19 years of soil management lead to the conclusion that minimal soil disturbance, coupled with crop residue retention, shaped autotrophic bacterial phylogenetics, modified soil physicochemical properties, and created a microenvironment that favored CO_2-fixing activity and increased soil productivity.

Keywords: carbon fixation; *cbbL* gene abundance; *cbbL*-harboring bacteria; carbon substrate utilization; soil management

1. Introduction

Soil microbial communities serve as the reservoir of biological processes and play a critical role in converting organic carbon into plant nutrients [1,2]. It is estimated that approximately 80 to 90% of the soil biogeochemical processes are mediated by microbial activities, including organic carbon degradation [3]. An essential biosynthetic process in soil is carbon dioxide (CO_2) fixation by autotrophic bacteria, a key process of carbon cycling [4] which is responsible for the sequestration of atmospheric CO_2 into the soil [5]. In agroecosystems, autotrophic CO_2-fixing bacteria are present in various types, including distinct physiological groups, such as cyanobacteria and purple photosynthetic and chemoautotrophic bacteria [6]. The groups of bacteria can incorporate ^{14}C into their microbial biomass through a series of biochemical reactions [7,8]. Those reactions are related to the community composition of autotrophic bacteria [9], influencing the intensity of CO_2 fixation.

The Calvin–Benson–Bassham cycle (i.e., the CBB cycle) is the most important autotrophic CO_2 fixation pathway [10,11] which influences the primary carbon source in an ecosystem [12]. Ribulose-1,5-bisphosphate carboxylase/oxygenase (known as RubisCO), a key enzyme in the CBB cycle, regulates the process of CO_2 reduction and the oxygenolysis of ribulose-1,5-bisphosphate [10,11]. RubisCO exists in four forms (referred as I, II, III, and IV) with different structures, catalytic activities, and O_2 sensitivities, of which form I's RubisCO mainly occurs in plant and photo- and chemo-autotrophic bacteria [13,14]. In the plant–soil–microbiome environment, autotrophic microbial diversity can be reflected by the functional biomarkers (i.e., the *cbbL* genes). These genes encode the large subunit of RubisCO form I that has a large scale of sequences for phylogenetic analysis [15,16].

Many anthropogenic activities, such as tillage and crop rotation, affect the composition and diversity of the microbial community in soil [17,18]. Soil tillage affects soil water conservation [19], fertility enrichment [20], and carbon sequestration [21,22]. Reduced or zero tillage (i.e., minimal to no disturbance to the soil profile) coupled with crop residue retention can increase the nutrient supply potential [23] and enhance carbon cycling [24] while decreasing the carbon footprint [25,26], which helps alleviate the challenge of global climate change [27]. Diversified crop rotation increases soil carbon [28,29], improves microbial diversity [30], and enhances the system's productivity and resiliency. However, little is known about the long-term impact of multiple years of soil tillage and cereal–legume alternate rotation on the function of autotrophic CO_2-fixing bacteria. It is unclear how the nexus of autotrophic bacterial community composition—CO_2 fixation—tillage practices impact soil microenvironments.

To elucidate those effects, we established the field experiment in 2001 on the western Loess Plateau of China. Different tillage treatments have since been applied to soil that has a cereal–legume crop rotation system. After the 19-year treatments (in 2019), soil was sampled from each plot in three replicates and analyzed for its biophysiochemical properties. The abundance and community composition of autotrophic CO_2-fixing bacterial communities were analyzed using real-time quantitative PCR and high-throughput sequencing. Soil carbon substrate utilization and its relation to autotrophic bacterial community were also determined. In a previous paper [31], we reported the diversity of bacterial communities and the characterization of the phylogenic composition in relation to soil management. In the present paper, we present the findings on (1) the changes in soil physicochemical parameters, autotrophic bacterial community composition, and soil catabolic diversity in response to the 19-year tillage treatments, and (2) the possible mechanisms responsible for shifts in the soil's autotrophic bacterial communities in relation to tillage. These findings from the long-term field study bring us to a new level of understanding of the interaction among autotrophic bacterial community, carbon fixation, and soil management practices.

2. Materials and Methods

2.1. Site Description

A field experiment was started in 2001 at the Rainfed Agricultural Education Center (35°28′ N, 104°44′ E; 1971 m a.s.l.) of Gansu Agricultural University in China. The site is on the western Loess Plateau and has a temperate, semiarid, continental monsoon climate [32]. Its solar radiation is 5.68 KJ m^{-2} and sunshine duration is 2476 h annually [33]. The mean temperature is 6.5 °C and the 10 °C-based accumulative temperature is 2339 °C. The average precipitation was 392 mm per year, with 54% occurring from June to September in most years. The experimental plots were on level terrain with dark loessal Calcaric Cambisol soil (FAO/UNESCO soil classification system, 1990). Prior to the start of the experiment in 2001, the site had a long history of the continuous cropping of flax (*Linum usitatissimum* L.) under conventional tillage practices [34].

2.2. Experimental Design

A randomized complete block design was used to accommodate the four soil management treatments in each of the three replicates. The description of the treatments and the implementation of the tillage practices and crop residue retention are summarized in Table 1. More details on the tillage and field operations of the treatments can be viewed in a previous publication [34]. In brief, all the treatments were implemented in both the wheat (*Triticum aestivum* L.) and field pea (*Pisum arvense* L.) phases of the rotation each year. For the TS and T treatments, the soil was plowed to a depth of 0.20 m in the fall and harrowed in the following spring's seeding time. The experiment included four tillage and crop residue management treatments per spring. In each of the three replicates, the following treatments were arranged in plot sizes of 80 m^2 (4 m wide × 20 m long). The crops were managed according to the recommended agronomic practices. The crops were planted in late March to early April, varying slightly from one year to the other due to weather conditions. The seeding rate was 187.5 kg ha^{-1} for wheat and 180 kg ha^{-1} for pea, on average, varying slightly each year due to seed size and percent germination. The seeding rates were aimed to achieve a plant population of 250 plants per m^{-2} for wheat and 80 plants per m^{-2} for field pea. The wheat was fertilized with urea nitrogen (46% N) at 105 kg N ha^{-1} and the pea at 20 kg N ha^{-1}. Both crops received a phosphorus fertilizer (Ca(H$_2$PO$_4$)2.2H$_2$O, with 16% P$_2$O$_5$) at 46 kg P$_2$O$_5$ ha^{-1}. The fertilizers were broadcast at sowing. No potassium (K) was applied as the soil contained exchangeable K greater than 300 mg kg^{-1}. No herbicides were applied to any crop because of the dry weather and little weed pressure. No irrigation was applied. At maturity (usually in mid-July to early August), the entire plot was harvested using a plot harvester and the biomass and yields were recorded.

Table 1. Treatment structure and the description of the tillage practices and crop residue implementation in the long-term field experiment started in 2001.

Treatment Abbreviation [a]	Tillage	Crop Residue
T	Conventional tillage, plowed to the 20 cm depth in the fall and harrowed the following spring	Crop residue was removed out of the field at harvest
NT	No tillage	Crop residue was removed out of the field at harvest
TS	Conventional tillage, plowed to the 20 cm depth in the fall and harrowed the following spring	Crop straw was chopped and incorporated in the soil via the fall plowing
NTS	No tillage	Crop straw was chopped, spread, and remained on the soil's surface

[a] The treatment descriptions and abbreviations will be used in the other Tables and Figures throughout the article.

2.3. Soil Sampling and Physicochemical Property Measurement

Bulk soils were sampled in June 2019 when pea plants were at the mid-flowering stage and wheat plants were at the late stage of flowering. During flowering, crop plants are most active, with roots interacting with soil microbes under semiarid environments [35,36]. Sampling at this stage can provide researchers with invaluable information about rhizospheres and their interactions with environmental factors [37]. From each plot, five soil samples were randomly taken from a 0 to 0.2 m depth using a 50 mm diameter iron soil corer, and the five samples were bulked for each plot. After passing through a 2 mm mesh sieve, two subsamples were taken from each plot: one was for DNA extraction and carbon substrate utilization assessment and the other was for soil property analysis.

The soil pH was determined with a pH meter (Mettler Toledo FE20, Shanghai, China) using a deionized suspension with 1:2.5 soil: water ratio (mass: volume). Soil organic carbon (SOC) was analyzed using the Walkley–Black wet oxidation method and total N (TN) was measured with the Kjeldahl method. Olsen phosphorus was examined using the colorimetric method with 0.5 M NaHCO$_3$. A UV-1800 spectrophotometer (Mapada

Instruments, Shanghai, China) was used to determine soil NH_4^+-N and NO_3^--N. Soil microbial biomass carbon and nitrogen (SMBC and SMBN, respectively) were measured using the Chloroform method. In each plot, soil moisture was measured gravimetrically and converted into volumetric water content using soil bulk density, which was measured by sampling the intact core of the soil using known-volume metal rings, drying the soil, and weighing the dried soil.

2.4. Soil Microbial Catabolic Diversity

For each soil sample from each plot (i.e., three replicates), community-level physiological profiles (CLPP) were determined using Biolog Eco-PlatesTM (Biolog Inc., Hayward, CA, USA) which contained 31 widely used C substrates along with a control well. The 31 C sources included nine carboxylic acids, seven carbohydrates, three other substrates, six amino acids, and two amines/amides [38]. Each well of the Eco-plates was inoculated with 150 µL of soil suspensions before being incubated at 25 °C. The carbon substrate utilization based on well color development was measured with optical density (OD) values at a wavelength of 590 nm (color development plus turbidity) and 750 nm (turbidity only) every 24 h for 7 d using the Biolog method (Biolog Inc., Hayward, CA, USA). Average well-color development (AWCD) was used to assess the microbial metabolic activity, as follows [39]:

$$\mathrm{AWCD} = \sum \frac{C - R}{31},$$

where C is the OD value of each well and R is the absorbance reading of the control well. Substrate richness and evenness, defined as the Shannon–Weaver index (H'), were calculated as follows:

$$H' = -\sum\nolimits_{i=n} p_i(lnp_i),$$

where *pi* represents the ratio of color development of a certain well to the sum of the color development of all wells in a micro-plate.

2.5. DNA Extraction and Quantification of cbbL Gene Abundance

Soil microbial DNA was extracted from each approximately 0.5 g of fresh soil sample, in triplicate, using a DNeasy PowerSoil Kit (QIAGEN, Inc., Hilden, Germany) following the manufacturer's protocol. The concentrations of the extracted DNA were determined with a NanoDrop ND-1000 spectrophotometer (Thermo Fisher Scientific, Waltham, MA, USA) and the quality of the extracted DNA was determined using agarose gel electrophoresis.

The *cbbL* gene copy number was assessed using real-time quantitative PCR (qPCR) on a LightCycler 480 II real-time PCR system (Roche Diagnostics, Mannheim, Germany). The *cbbL* genes were amplified with the forward primer K2f (5′-ACCAYCAAGCCSAAGCTSGG-3′) and the reverse primer V2r (5′-GCCTTCSAGCTTGCCSACCRC-3′) (reference). The reaction mixture contained 1 µL of DNA template, 5 µL of 2 × LightCycler® 480 SYBR Green I Master, 0.2 µL of forward primer, 0.2 µL of reverse primer, and 3.6 µL of nuclease-free water, and it made up a final volume of 10 µL. The thermal protocol was 10 min of incubation at 95 °C, followed by 40 cycles of 10 s at 95 °C and 30 s at 60 °C. Melting curve analysis was performed to validate the specificity of the amplified PCR products. The standard curves were produced using triplicate 10-fold dilutions of plasmid DNA with inserted *cbbL* gene. PCR amplification efficiencies were 94% with an R^2 value of 0.99.

2.6. High-Throughput Sequencing of the cbbL Gene

The primer set (K2f-V2r) for *cbbL* gene real-time qPCR was used for the *cbbL* gene amplification for each sample, respectively. The PCR components contained 5 µL of Q5 reaction buffer (5×), 5 µL of Q5 High-Fidelity GC buffer (5×), 0.25 µL of Q5 High-Fidelity DNA polymerase (5 U/µL), 2 µL (2.5 mM) of dNTPs, 1 µL (10 uM) of each forward and reverse primer, 2 µL of DNA template, and 8.75 µL of ddH$_2$O. The PCRs were performed with the following procedures: 2 min of initial denaturation at 98 °C, followed by 25 cycles of 15 s at 98 °C, 30 s for annealing at 55 °C, 30 s for elongation at 72 °C, and a final extension at 72 °C for 5 min. PCR amplicons were extracted from 1% agarose gels and purified with Agencourt AMpure beads (Beckham Coulter, Indianapolis, IN, USA) and quantified using the PicoGreen dsDNA assay kit (Invitrogen, Carlsbad, CA, USA). High-throughput sequencing data were obtained through commercial laboratory services (Personal Biotechnology Co., Shanghai, China) and deposited in the NCBI database with accession PRJNA689959.

2.7. Data Processing and Bioinformatics Analysis

Raw fastq files were quality-filtered and merged using VSEARCH pipeline (version 2.13.4) on the platform of QIIME 2. The sequences with a quality score <20, having ambiguous bases, or those containing mononucleotides repeats of >8 bp were removed. The operational taxonomic units (OTUs) were clustered with a 97% sequence similarity cut-off using VSEARCH pipeline (v. 2.13.4). All representative reads in the OTUs were taxonomically classified and annotated based on the GenBank® nucleotide sequence database (http://ncbi.nlm.nih.gov; accessed on 15 March 2020).

2.8. Statistical Analysis

Data were analyzed using SPSS software (SPSS Inc., Chicago, IL, USA). Significant differences between treatments in soil properties, *cbbL* gene abundance, and catabolic diversity were determined at the probability of <0.05.

The QIIME2 package was employed to determine alpha diversity, and the community evenness and richness were expressed using the Shannon and Simpson indexes. The variation of *cbbL*-harboring bacterial community composition across the samples was determined using principal co-ordinates analysis (PCoA) based on the genus-level compositional profiles. Hierarchical clustering analysis was performed with the unweighted pair-group method of arithmetic means (UPGMA) based on Bray–Curtis distance matrices. The differences in the taxonomic and phylogenetic communities were compared by clustering them using the R *"vegan"* package (R Foundation for Statistical Computing, Vienna, Austria). The significant differences in microbial structure among treatments was determined by permutational multivariate analysis of variance (PERMANOVA) and analysis of similarities (ANOSIM).

Network analysis was performed to investigate the co-occurrence patterns among the *cbbL*-harboring bacterial communities. The OTUs with a relative abundance of higher than 0.01% were retained and the *"psych"* package of R version 4.0.2 was used to analyze the preprocessed data and calculate the spearman correlation coefficient matrix. Statistical correlations with a cut-off at an absolute *r* value of higher than 0.6 and a *p* value of lower than 0.05 were retained for further analysis. The software Gephi (version 0.9.2) was employed for network visualization and network properties measurement. To identify the keystone bacterial taxa driving community turnover under different treatments, the abundance of bacterial taxonomic OTUs was regressed against the four field treatments based on random forest classification analysis using the *"randomForest"* package of R version 4.0.2. Lists of biomarkers taxa were ranked to compare the importance of OTUs using 10-fold cross-validation, implemented using R *'rfcv'* function. The top 15 bacterial biomarkers were chosen according to the stabilized cross-validation error curve. The relative abundance of biomarker taxa was illustrated using the *"heatmap"* package of R version 4.0.2. The Spearman's correlation coefficients and the Mantel test with 999 permutations in the *"vegan"* package of R were used to assess the correlations of soil properties and the distribution of the dominant genera of *cbbL*-harboring bacteria based on the distance matrix.

3. Results

3.1. Soil Physiochemical and Biological Properties

The soil's biophysiochemical properties significantly differed between treatments (Table 2). The contents of SOC, TN, TK, SMBC, and SMBN were significantly greater for the NT and NTS treatments compared to the T and TS treatments ($p < 0.05$). Soil organic carbon was 3.2%, 13.4%, and 6.1% greater for the NT, NTS, and TS treatments, respectively, compared with the T treatment. The highest soil TN concentration (1.07 g kg^{-1}) was obtained in the NTS soil. The TK for the NTS treatment was 10.9% ($p < 0.05$) greater than that for the NT, TS, and T treatments. The soil's bulk density was 4.6%, 9.2%, and 6.1% lower for the NT, NTS, and TS treatments, respectively, compared with the T treatment. The NT, NTS, and TS treatments increased SMBC by 12.6%, 31.8%, and 17.0%, respectively, compared to the T treatment. In addition, SMBN for the NTS treatment was greater ($p < 0.05$) than that for the T treatment ($p < 0.05$), with no difference among NT, TS, and T. There was no difference in soil pH, TP, NO_3^--N, NH_4^+-N, soil temperature, and moisture among the treatments ($p > 0.05$).

Table 2. Biophysiochemical properties of the soils under different management practices.

Soil Parameter [a]	Treatment [b]			
	NT	NTS	T	TS
pH (H_2O)	8.13 ± 0.03 a [c]	8.10 ± 0.03 a	8.13 ± 0.02 a	8.14 ± 0.07 a
SOC (g kg^{-1})	12.77 ± 0.61 ab	14.03 ± 0.40 a	12.37 ± 0.81 b	13.13 ± 0.60 ab
TN (g kg^{-1})	0.86 ± 0.07 ab	1.07 ± 0.13 a	0.78 ± 0.08 b	0.88 ± 0.09 ab
TP (g kg^{-1})	0.56 ± 0.02 a	0.66 ± 0.16 a	0.54 ± 0.06 a	0.55 ± 0.05 a
TK (g kg^{-1})	18.61 ± 0.90 ab	19.29 ± 0.44 a	17.39 ± 0.77 b	18.86 ± 0.67 ab
Olsen p (mg kg^{-1})	13.43 ± 2.71 a	15.89 ± 1.95 a	14.04 ± 4.30 a	15.38 ± 1.03 a
NO_3^--N (mg kg^{-1})	37.54 ± 3.38 a	35.79 ± 1.02 a	36.26 ± 3.95 a	33.84 ± 3.20 a
NH_4^+-N (mg kg^{-1})	1.28 ± 0.29 a	1.48 ± 0.09 a	1.40 ± 0.13 a	1.20 ± 0.13 a
Moisture (%)	13.57 ± 0.43 a	14.76 ± 0.97 a	13.56 ± 0.35 a	13.87 ± 0.44 a
Bulk density (g cm^{-3})	1.25 ± 0.03 b	1.19 ± 0.02 c	1.31 ± 0.02 a	1.23 ± 0.03 bc
Total porosity (%)	53.05 ± 0.28 a	53.52 ± 0.79 a	50.43 ± 0.72 b	51.25 ± 0.32 b
SC (mm h^{-1})	77.16 ± 2.24 b	86.28 ± 2.09 a	73.05 ± 0.53 c	86.21 ± 1.45 a
Temperature (°C)	23.20 ± 0.96 a	23.63 ± 0.70 a	23.10 ± 0.66 a	24.03 ± 0.55 a
SMBC (mg C kg^{-1})	185.51 ± 8.82 ab	217.12 ± 11.35 a	164.69 ± 12.31 b	192.75 ± 16.92 ab
SMBN (mg N kg^{-1})	19.27 ± 0.66 b	22.01 ± 1.61 a	18.22 ± 0.61 b	18.20 ± 0.53 b

[a] SOC, soil organic carbon; TN, total nitrogen; TP, total phosphorus; TK, total potassium; SC, saturation conductivity; SMBC, soil microbial biomass carbon; SMBN, soil microbial biomass nitrogen. [b] The treatment descriptions are summarized in Table 1. [c] Data (means ± SD, n = 3) in each row followed by different letters are significantly different at $p < 0.05$.

3.2. cbbL Gene Abundance and the Relation to Soil Properties

The soils under the NTS treatment had more copy numbers of *cbbL* genes than the soils under the other treatments, and the increment was 52% when NTS was compared with the T treatment ($p < 0.01$; Table 3). Correlation analysis across the treatments showed that the soil *cbbL* gene was positively correlated with soil moisture, saturation conductivity, SMBC, and SMBN, but it was negatively correlated with bulk density ($p < 0.05$; Table S1).

Table 3. The *cbbL* gene copy numbers of the soil samples under the different soil management practices.

Treatment Abbreviation [a]	*cbbL* Abundance (10^8 Copies g^{-1} Soil)
NT	2.17 ± 0.42 b [b]
NTS	3.51 ± 0.36 a
T	2.31 ± 0.30 b
TS	2.39 ± 0.27 b

[a] The treatment descriptions are summarized in Table 1. [b] Values in each column followed by different letters are significantly different at $p < 0.05$.

3.3. Autotrophic Bacterial Community Diversity and OTUs Richness

After quality control, a total of 468,535 filtered sequences with 32,250 to 43,646 sequences per sample remained (Table S2). The rarefaction curves were close to the saturation phase with the increase in sample size. There was sufficient sequencing depth and the OTUs were representative of the overall bacterial community libraries. To compare soil carbon fixation and microbial community diversity, the same survey effort level of 30,637 sequences was randomly selected from each soil sample (95% of the smallest number of reads). In total, 4988 OTUs were observed across all samples, with the number of OTUs ranging from 503 to 1692, varying with the soil samples (Table S2). The four tillage and crop residue management treatments shared 451 OTUs, accounting for 9.04% of the core OTUs (Figure 1a). There were differences in the OTU-sharing between treatments: the NTS and NT treatments shared the highest number of OTUs (1107 OTUs, 22.19%), whereas the TS and T treatments shared the least number of OTUs (683 OTUs, 13.69%) and the NTS and T treatments shared 18.04% of the OTUs (900 OTUs). Moreover, the NTS treatment harbored the highest number of unique OTUs (1458 OTUs, 29.23%) and the T treatment harbored the lowest number of unique OTUs (308 OTUs, 6.17%).

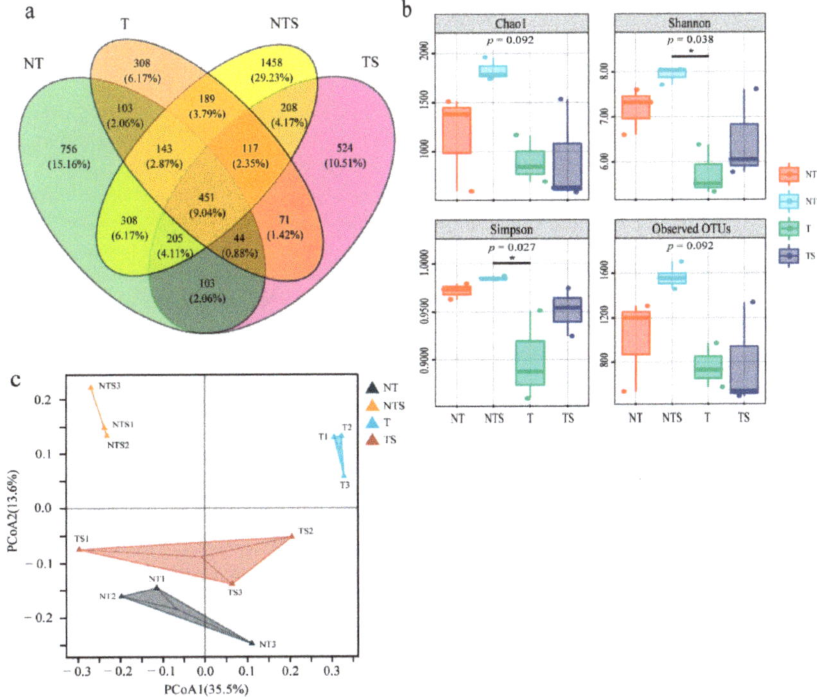

Figure 1. OTUs richness and diversity of soil CO_2-fixing bacteria communities across all the treatments. (**a**) Venn diagram containing the unique and shared and unique OTUs among the treatments. (**b**) Alpha diversity index including Chao1, Simpson, Shannon, and observed OTUs. The asterisks indicate significant differences among the treatments. (**c**) Principal coordinate analysis (PCoA) based on bacterial community composition. The treatment descriptions are summarized in Table 1. The numbers following the treatment name denote the sampling replications. For example, NT1, NT2, and NT3 mean the soil sampling, respectively, from replicate 1, 2, and 3 of the NT plots.

The alpha diversity of bacterial populations differed significantly among soil treatments, as shown by the Chao1, Simpson, Shannon, and observed OTUs indexes ($p < 0.05$; Figure 1b). The Shannon index for the NTS treatment was significantly higher compared to the T treatment and the diversity was ranked NTS > NT > TS > T. No differences in the Chao1 and observed OTUs indexes were observed among the four treatments ($p > 0.05$; Figure 1b). The PCoA results showed similarities and differences in CO_2-fixing bacterial community structure across the different treatments (Figure 1c), where total variance explained by PCoA1 and PCoA2 was 35.5 and 13.6%, respectively. The bacterial communities of the soil under the NT and TS treatments were clustered closely, whereas the community compositions in the soil under the NTS and T treatments differed from those of the NT and TS treatments. Permutational multivariate analysis of variance (PERMANOVA) and analysis of similarities (ANOSIM) confirmed the significant differences in the compositions of the soil *cbbL*-carrying bacterial communities under different tillage practices and crop residue managements ($p < 0.05$; Table S3).

3.4. Composition and Networks of Soil Autotrophic Bacteria Communities

The composition of the bacterial communities was determined for soils from each of the three field replicates (e.g., T1, T2, and T3 for the T treatments in Figure 2 and Table S4). At the phylum level, the bacterial communities were dominated primarily by the phyla *Proteobacteria* (38.52–80.94%) and *Actinobacteria* (5.72–22.65%), both accounting for an average of 56–85% of the relative abundance of the bacterial communities (Figure 2a; Table S4). The phyla *Euryarchaeota* and *Cyanobacteria* had low relative abundances. The tillage treatment had a significant ($p < 0.05$) impact on the relative abundance of a bacterial community. The relative abundance of the phylum *Proteobacteria* was 34.2, 41.6, and 33.5% lower in the soil under NT, NTS, and TS, respectively, while the relative abundance of the phylum *Actinobacteria* was 158.4, 245.7, and 142.5% higher, respectively, than that measured in the soil under the T treatment.

At the class level, *alpha-Proteobacteria* and *beta-Proteobacteria* dominated the bacterial communities, containing 14.89–30.71% and 5.71–22.65% of the total *cbbL* gene sequences, respectively (Figure 2b; Table S4). The class *Rubrobacteria* was the most dominant within the phylum *Actinobacteria*, which accounted for 0.57–4.00% of the total relative abundance. At the genus level, the *cbbL*-carrying bacteria were mainly affiliated with *Bradyrhizobium*, *Variovorax*, *Gaiella*, and *Steroidobacter*, accounting for 20–35% of the identified communities (Figure 2c; Table S4). The tillage treatments had an impact on the community composition, as shown in the top 10 genera; compared to the T treatment, the NTS treatment increased the relative abundance of genus *Bradyrhizobium* by 59.3% and reduced the relative abundance of genus *Variovorax* by 60.7%. However, no significant differences in the taxonomic composition were observed among the NT, TS, and T treatments. The hierarchical cluster analysis (Figure 2c) had results similar to those obtained from PERMANOVA and ANOSIM (Table S4), where the bacterial communities formed two main clusters (T and TS vs. NT and NTS), and the largest differences were between the NTS and T treatments.

Co-occurrence network analyses were conducted to evaluate the different co-occurrence patterns of the *cbbL*-harboring bacterial communities under the different treatments (Figure 3). In total, there were 35 nodes and 71 links, with an average degree of 4.057 and a graph density of 0.119 in the autotrophic bacterial network (Table S5). The ratio of positive edges was higher than the ratio of negative edges in the *cbbL*-harboring bacterial communities, with the ratio of positive edges accounting for 85.92%. The genera *Methylocella* and *Advenella* were the most connected keystone genera in the *cbbL*-harboring bacterial communities, and both were from the phylum *Proteobacteria*, with significant correlations with other genera (Figure 3; Table S5).

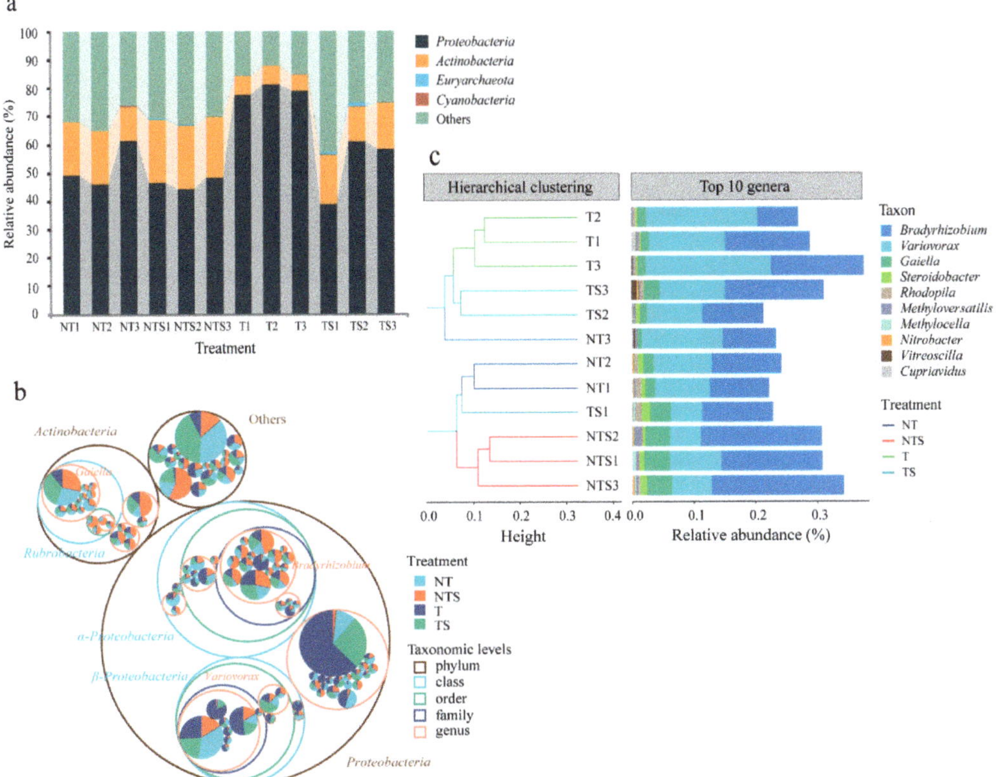

Figure 2. Microbial community composition under the four tillage practices and crop residue management treatments. (**a**) Relative abundance (%) of the most abundant phylum across all soil samples. (**b**) Taxonomic differences of *cbbL*-harboring microbial community. The largest circles represent the phylum level and the inner circles represent class, order, family, and genus. (**c**) Hierarchical cluster analysis and relative abundance (%) of the top 10 *cbbL*-harboring bacterial genera. The treatment descriptions are summarized in Table 1. The numbers following the treatment name denote the sampling replications. For example, NT1, NT2, and NT3 mean the soil sampling, respectively, from the NT plots in each of the three replicates.

3.5. Biomarkers in the Autotrophic Bacterial Communities

To determine the key biomarkers discriminating the *cbbL*-harboring bacterial communities under different treatments, we performed a regression of the abundances of bacterial OTUs against the tillage treatments based on the random forests machine learning algorithm. The top 15 keystone OTUs were chosen as the respective bio-marker taxa because the cross-validation error curve was stabilized when using these OTUs for both bacterial communities (Figure 4). The biomarker taxa were mainly affiliated with the classes of *alpha-Proteobacteria* and *beta-Proteobacteria* in the *cbbL*-harboring bacterial communities (Table S6). The heatmaps in Figure 4 further illustrate the scaled abundances of these biomarker taxa in response to soil tillage management. It was found that the phylum *Proteobacteria* played more crucial roles than the other phyla across soil samples for the soil's *cbbL*-harboring bacterial communities.

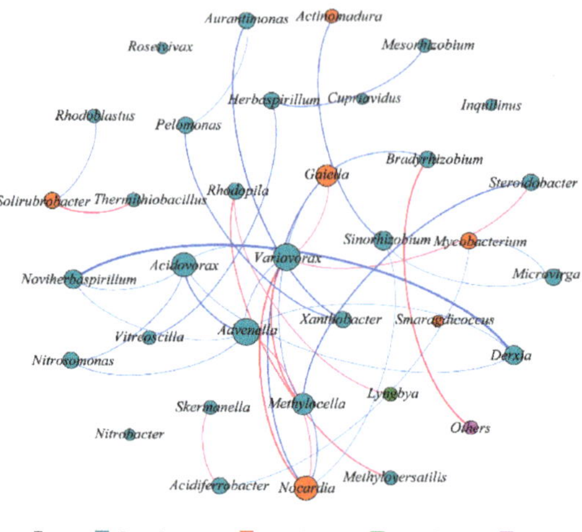

Figure 3. Spearman correlation analysis showing the *cbbL*-harboring bacterial co-occurrence networks across the soil samples. Each connection line represents a statistically significant correlation ($p < 0.05$) with a Spearman correlation coefficient of $r > 0.6$. The nodes of the networks are colored at the phylum level and labeled at the genus level. The size of the node is proportional to the number of connections (degrees), and the thickness of the edge is proportional to the value of the Spearman's correlation coefficients. The blue edges stand for positive correlations between two bacterial nodes, while the red edges stand for negative correlations.

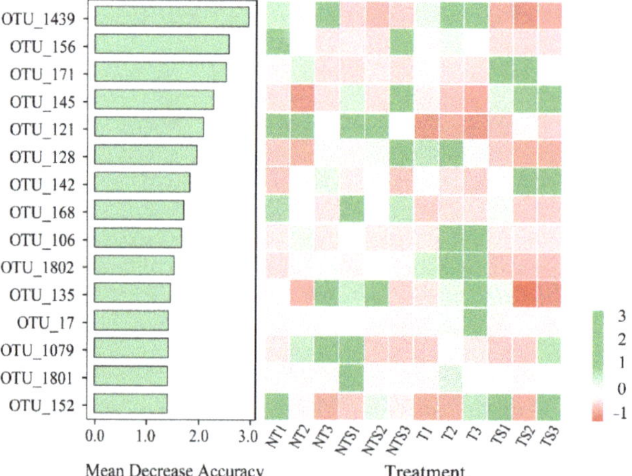

Figure 4. Bacterial taxonomic biomarkers in *cbbL*-carrying bacterial communities. The top 15 biomarker taxa were ranked in descending order of the importance to the accuracy of the models. The heatmap illustrates the variation in the abundance of the 15 predictive biomarker taxa (OTUs). The treatment descriptions are summarized in Table 1. The numbers following each treatment name in the heatmaps denote the sampling replications. For example, NT1, NT2, and NT3 mean the soil samples were taken from replicate 1, 2, and 3 of the NT plots, respectively.

3.6. Soil Microbial Catabolic Diversity and Carbon Utilization Pattern

The soil's microbial metabolic activity was expressed as the average well-color development (AWCD), an indicative measurement of the physiological profiles at the community level. The AWCD differed significantly between the treatments (Table S7). The NTS and TS treatments increased the AWCD index by 21.6 and 19.1%, respectively, while the NT treatment reduced the AWCD index by 15.4% in comparison with the T treatment. However, no significant differences for the *H′* (Shannon–Wiener) index were observed among the four treatments.

The soil's microbial carbon utilization patterns based on six fundamental groups differed among the treatments (Figure 5a). Compared to the T treatment, NT and TS reduced carbon resource utilization based on amines/amides by 63.6 and 35.2%, respectively, while NTS and TS increased carboxylic acid utilization by 37.1 and 44.9%, respectively. The overall carbon utilization capabilities differed significantly among the six carbon substrate groups (Figure 5b). The carbon utilization for carbohydrates, carboxylic acids, and amino acids were significantly higher than that for the polymers, miscellaneous, and amines/amides groups. The highest carbon utilization was obtained in the carbohydrates group, with a mean value of 8.808, while the lowest carbon utilization was found in the amines/amides group, with a mean value of 0.815.

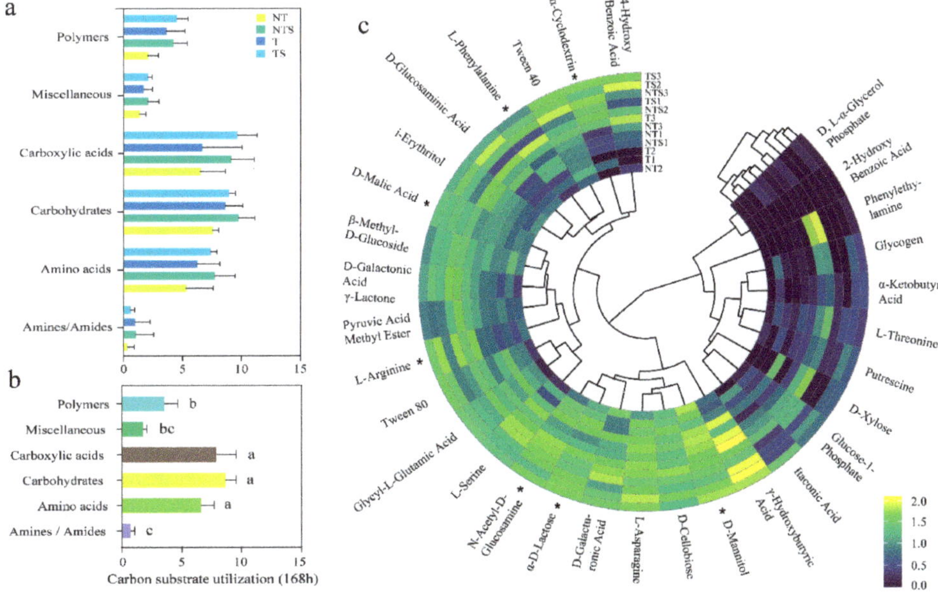

Figure 5. Carbon utilization patterns of soil bacterial communities under the four tillage and crop residue management treatments, with (**a**) the carbon utilization based on six carbon substrate groups, (**b**) the carbon utilization of six carbon substrate groups, and (**c**) the utilization capabilities of 31 carbon substrate resources among the treatments. The different lowercase letters in (**b**) indicate significant differences among the substrates at $p < 0.05$. The asterisks in (**c**) indicate significant differences among the tillage treatments. Data are means ± standard deviation (n = 3). The treatment descriptions are detailed in Table 1.

The carbon utilization capability of soil microbiomes, based on seven specific substrates (L-Arginine, L-Phenylalanine, α-D-Lactose, N-Acetyl-D-Glucosamine, D-Mannitol, D-Malic Acid, and α-Cyclodextrin), differed significantly among the tillage treatments (Figure 5c). The utilization of L-Arginine substrate in the TS treatment was 73.9% higher ($p < 0.05$) than that in the T treatment. The NTS treatment increased the utilization of L-Phenylalanine substrate by 85.5%, compared with the T treatment ($p < 0.05$). The NT treatment reduced the utilization of α-D-Lactose and N-Acetyl-D-Glucosamine substrates by 28.9 and 34.4%, respectively, compared with the T treatment ($p < 0.05$). The NT treatment increased the utilization of D-Mannitol substrate by 18.67% ($p < 0.05$) compared with the NTS, TS, and T treatments, with the latter three having a similar utilization. For the utilization of D-Malic Acid substrate, the NTS and TS treatments respectively had 38.8 and 45.6% greater ($p < 0.05$) utilization than the T treatment. The utilization of α-Cyclodextrin substrate was 68.9% greater in the NTS and 102.7% greater in the TS treatments compared to the T treatment, while the NT treatment reduced the α-Cyclodextrin substrate utilization by 64.8% compared with the T treatment ($p < 0.05$). Across the four tillage treatments, the mean carbon utilization capabilities of the soil microbiomes differed significantly among the 31 carbon substrates. D-Mannitol substrate had the highest carbon utilization, averaging 1.624, and 2-Hydroxy benzoic Acid substrate had the lowest utilization, averaging 0.023.

3.7. Correlation among Physiochemical Traits, Catabolic Diversity, and the Autotrophic Bacterial Community

There were significant ($p < 0.05$) correlations among the soil physicochemical properties and *cbbL*-harboring bacterial community compositions (Figure 6a). Spearman correlation analysis across the tillage treatments revealed that SOC was positively correlated with TN, saturation conductivity, SMBC, and SMBN, but it was negatively correlated with bulk density (Table S8). Total soil N was correlated positively with TK, saturation conductivity, and SMBN, while soil bulk density was negatively correlated with saturation conductivity, SMBC, and SMBN. Further, the Mantel test revealed that the structure of the autotrophic bacterial communities was highly correlated with SOC ($r = 0.315$, $p < 0.05$), TK ($r = 0.328$, $p < 0.05$), bulk density ($r = 0.724$, $p < 0.05$), total porosity ($r = 0.415$, $p < 0.05$), saturation conductivity ($r = 0.349$, $p < 0.05$), and SMBC ($r = 0.377$, $p < 0.05$) (Figure 6a).

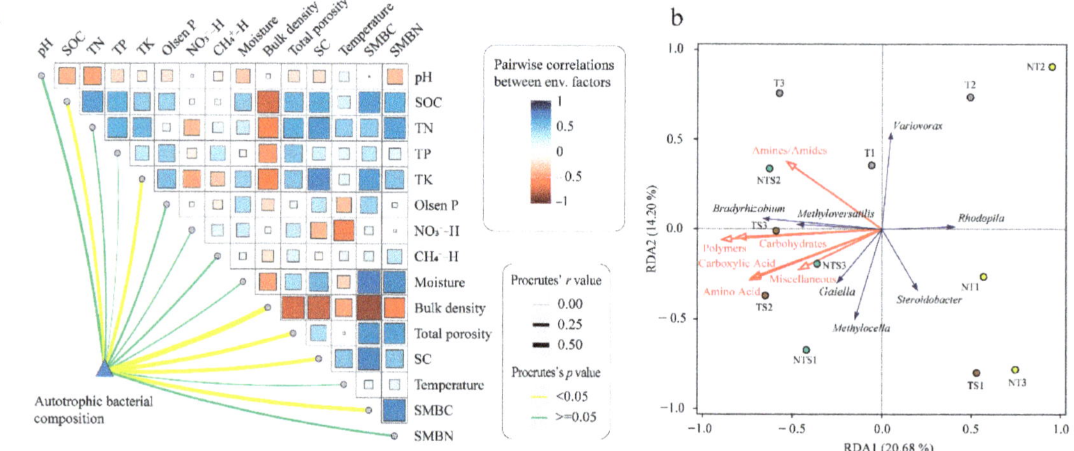

Figure 6. A complex of correlations (**a**) between the autotrophic bacterial compositions (Table S4) and environmental factors (including soil physiochemical properties) and (**b**) between autotrophic bacterial community and microbial catabolic diversity in response to soil tillage treatments. In the (**a**) analysis, the pairwise comparisons of environmental factors are shown with the color gradient denoting Spearman's correlation coefficients (range 1 to −1); Spearman's rank correlations (r) of key soil parameters; and Mantel tests for the correlation between environmental factors and the distribution of the autotrophic bacterial community compositions. The edge width in the connection lines denotes the Mantel's r statistic for the corresponding distance correlations, and the edge color represents the statistical significances based on 999 permutations. In the (**b**) analysis, the red lines represent carbon substrate utilization, the blue lines represent the bacterial genus-level taxonomy, and the different colored circles represent the soil samples from all replications (n = 3) of each treatment. The treatment descriptions are summarized in Table 1.

Redundancy analysis, showing the relationship between carbon source utilization and the structure of *cbbL*-harboring communities at the genus level, revealed that the community composition varied from 14.20 to 20.68% (Figure 6b). The carbon utilization of polymers, amino acids, carbohydrates, and carboxylic acid were clustered to the edge of the genus *Bradyrhizobium* and *Methyloversatilis*, and they were negatively correlated to the genus *Rhodopila*. Among the carbon substrates, polymers explained 16.6% of the variance ($F = 2.0$, $p = 0.178$) and amino acids explained 8.1% of the variance ($F = 0.9$, $p = 0.414$) across the soil samples, and they were the key indicators of carbon utilization capacity (with 999 permutations) (Table S9). The abundance and diversity of the genus *Gaiella* were closely related to miscellaneous. The 19 years of tillage treatments had an impact on bacterial community structure, with the genera *Bradyrhizobium* and *Gaiella* presenting highly in the soil under the NTS treatment, while the genus *Variovorax* was high in the soil with continuous T practices.

4. Discussion

4.1. Diversity of Soil cbbL-Carrying Bacterial Communities Shaped by Crop Residue Retention

The biomarker *cbbL* genes, which encode a large subunit of ribulose-1,5-bisphosphate (RUbP) carboxylase/oxygenase in the Calvin–Benson–Bassham cycle, are considered important indicators of CO_2-fixing autotrophic microbes in soil. We found a significant shift in the phylogenetic diversity of *cbbL*-carrying bacteria in response to soil management, and the largest differences in diversity existed between the NTS and the T treatments. The NTS treatment reduced the relative abundance of *Proteobacteria* while increasing the abundance of *Actinobacteria* at the phylum level, and at the genus level, the NTS treatment increased the relative abundance of *Bradyrhizobium* while reducing the relative abundance of *Variovorax*. The mechanisms for the shift in the bacterial community composition due to different soil management treatments are unclear, but we suggest the following possible mechanisms: **(a)** the minimal physical disturbance to the soil, coupled with continuous crop residue input (in each growing season, year-after-year), provides cumulative feedback to the soil's microenvironment, forming a balanced C to N ratio over the years, which favors microbial activities. A high N availability in soil is found to promote *Proteobacteria* while prohibiting *Actinobacteria*. A high C availability in soil is found to stimulate *Proteobacteria* while prohibiting other microbes compared to low N or low C soils. **(b)** The 19 years of continuous input of crop residue as organic material in the NTS treatment increased C sources for energy, essential for the cell synthesis of organisms [17], leading to the increased abundance of *cbbL* genes encoding ribulose-1,5-bisphosphate carboxylase oxygenase for CO_2 fixation. The decomposition of the input material increased the substrate sources, leading to the great shift in the taxonomic composition of *cbbL*-carrying bacterial communities. **(c)** The shift in the microbial community composition due to soil management led to the formation of a microenvironment favorable to the humification process of SOC, increasing the size of the organic C pools. Additionally, **(d)** the facultative CO_2-fixing bacterial members in soil may grow heterotrophically with the addition of crop residue (besides autotrophic activities) [40,41]. Nevertheless, our results, in combination with the findings reported by others [42–44], bring us to a new level of understanding that long-term crop residue input with minimal disturbance can shape microbial community composition, benefiting CO_2 fixation processes and soil C cycling.

4.2. cbbL-Carrying Bacterial Catabolic Diversity in Relation to Soil Management

Bacterial OTU richness and catabolic diversity were highest in the soil under no till with crop residue retention among the soil treatments evaluated, suggesting that soil management does alter the diversity of soil microbiota, provided that the soil is treated for a longer period of time (19 years in the present study). Other researchers have found that the catabolic diversity is reflected by the shift in microbial community composition accompanied by the different capacities of carbon substrate utilization [45]. However, the catabolic diversity found in bulk soils may not be detectable in rhizosphere soils [46], since the expression of genes for soil nutrient reduction varies with soil microenvironments. Further, we found that the 19 year soil treatments did not affect soil pH (Table 1). The *cbbL*-carrying taxa were mainly affiliated with the types of bacteria, regardless of the level of pH tolerance. The diversity of autotrophic bacterial communities was highest in soil under the NTS treatment and lowest in soil under the T treatment, mainly due to the continuous lack of disturbance and crop residue input that created an ideal microenvironment that favored microbial activities. Our results support the findings of some researchers that soil pH does not have a direct role in impacting bacterial community diversity [47,48], and they do not agree with some others [49,50]. We suggest that the relationship between soil pH and bacterial community diversity is likely driven by pH-related soil nutrients. For example, the heavy use of N fertilizers can reduce the diversity and abundance of *cbbL* gene-containing microorganisms [51], mainly due to soil N decreasing pH, which prohibits some beneficial microbial activities [17,52].

4.3. Relationships among Physiochemical Properties, CO_2 Fixation, and Bacterial Diversity

Soil is a comprehensive ecosystem in which microbial communities have a strong correlation with soil management options, host plants, and edaphic factors [53,54], including nutrient availability, ion exchange capacity, and SOC turnover [55,56]. In our study, the 19 year soil management practices led to substantial changes in physiochemical and biological properties. No till coupled with crop residue retention significantly increased *cbbL* gene abundance, SOC, SMBC, and SMBN while decreasing bulk density compared to conventional tillage. Our results add new knowledge that in a no-till soil microenvironment, bulk density, SMBC, and SMBN impact the *cbbL* gene abundance, and that the increased *cbbL* gene abundance, in turn, increases CO_2 fixation, affirming the positive role of *cbbL*-carrying bacteria in soil's C cycling. The *cbbL*-carrying facultative autotrophs played a more prominent role than the obligate autotrophs in soil CO_2 fixation, as we identified both facultative and obligate bacteria through sequencing, and we found that the relative abundance of *Proteobacteria* at the phylum level was highest in all the soil treatments evaluated. *Proteobacteria* is a well-known facultative aerobic autotroph that assimilates CO_2 via the Calvin–Benson cycle. However, the CO_2-fixing capability of autotrophs may vary with other factors, such as plant species [57], crop fertilization [51], and soil amendments [58].

4.4. Implication of Long-Term Field Studies

Bacteria play an important role in soil C cycling, some of which serve as the decomposers of root exudates, plant litter, and other carbon sources which break down the organisms and convert organic energy into forms useful to the rest of the organisms in the food web. The conversion process releases nutrients and thus improves soil fertility. However, limited information is available regarding how the bacterial community composition and diversity are influenced by long-term soil management options, such as tillage practices and crop residue input. In the scientific literature, some aspects of the relationship between soil bacterial communities and anthropogenic activities are documented, for example, soil tillage practices shape microbial community composition [59,60], change microbial diversity [46,61], and provide feedback to soil physiochemical properties [62]. However, most published findings are primarily drawn from short-term (< 10 years) experiments, such that the conclusions are often inconclusive, inconsistent, or even controversial from one study to another. Running a large field experiment continuously, year after year, is undoubtedly costly, but the experience of conducting the 19-year field experiment has provided us with the high confidence for making concrete conclusions on the subject area.

The improved soil microenvironmental conditions due to no till and crop straw retention for 19 years has led to increased soil productivity. Pea yield ranged from 1030 to 1980 kg ha^{-1} and aboveground plant biomass ranged from 4300 to 6400 kg ha^{-1} (Figure S1). Spring wheat followed a similar trend of treatment effect as that obtained in field pea. The effect of soil management on crop yields in 2019 was similar to that averaged cross the 19 study years; the NTS, TS, and NT treatments increased ($p < 0.01$) yields by 32.5,

25.1, and 10.0%, on average, of wheat and pea, respectively, compared to the T treatment. The soils receiving residue retention treatments (i.e., NTS and TS) had significantly higher productivity (22.7% more) compared to the soil without crop residue retention (i.e., NT and T).

Knowledge of the complex nexuses among soil's physiochemical properties, microbiomes, and anthropogenic activities (such as tillage practices and crop residue input) is highly demanded in the development of strategies to improve the health of soil microenvironments. Our findings are well positioned to meet the needs. One of the important aspects in the assessment of the impact of long-term soil management practices is to quantify the relative weights or percent impact between the two subfactors—tillage practice and crop straw retention. There is a need to understand whether the combination of the two subfactors actually generates additional synergy between the two. For example, the NTS treatment in our study had the greatest benefits in terms of soil microbial diversity, C-cycling, and nutrient enrichment compared to the rest of the treatments; however, we were unable to differentiate the relational weight of the impact between tillage practice and straw retention. Did the benefits stem from the NTS treatment due to straw input (x% of the benefits), no tillage (y% of the benefits), or the combination of the two? Nevertheless, our long-term experiment has been running since 2001 at a site representative of the 1.56 million hectares of the China Loess Plateau. The results of the study can provide some guidelines for improving soil health and productivity in local regions, and possibly extended to areas with similar climatic conditions to the experimental site.

5. Conclusions

The results of the 19-year field experiment show that continuous disturbance to the soil profile elevated the selective pressure to the soil CO_2-fixing bacteria community, disturbed microbial catabolic activity, and created hardship for the microenvironment. Conversely, no disturbance to the soil, coupled with residue-retaining on the soil surface or crop residue being incorporated into the soil, increased the soil's organic carbon and nutrients and lowered the soil's bulk density. The underlying micro-driven functional changes led to enhanced soil carbon utilization. We conclude that bacterial community structure, composition, and CO_2-fixing capability are highly regulated by soil management practices, and that a minimal disturbance to the soil's microenvironment, coupled with the retention of crop residues in the soil, will improve bacteria-involved biological activities and increase nutrient cycling and soil productivity.

Supplementary Materials: The following are available online at https://www.mdpi.com/article/10.3390/agriculture12091415/s1, Table S1: Pearson correlation coefficients between *cbbL* gene abundance and soil physicochemical properties, Table S2: Detailed sequencing depth and OTUs number of the *cbbL*-harboring bacterial communities under different soil treatments, Table S3: PERMANOVA and ANOSIM results for the effects of tillage and crop residue management treatments on soil *cbbL*-harboring bacterial communities, Table S4: Relative abundance of soil *cbbL*-harboring bacterial taxonomic composition at different levels for all samples as affected by tillage and crop residue management treatments, Table S5: Co-occurrence network properties of soil *cbbL*-harboring bacterial communities under the four tillage and crop residue management treatments, Table S6: Lists of the top 15 bacterial biomarkers taxa of *cbbL*-harboring bacterial communities under the different treatments, Table S7: Soil microbial catabolic diversity response to the four tillage and crop residue management treatments, Table S8: Spearman correlation coefficients among the soil's physicochemical and biological properties, and Table S9: Redundancy analysis of the relationship between carbon source utilization and the *cbbL*-harboring bacterial community at the genus level and clustering of the soil samples. Figure S1: Annual biomass and grain yield of spring wheat and field pea grown under different soil management practices.

Author Contributions: Conceptualization, L.L.; validation, C.L.; resources, L.L.; writing—original draft preparation, C.L.; writing—review and editing, L.L.; supervision, L.L.; project administration, J.X., L.C. and Z.L. All authors have read and agreed to the published version of the manuscript.

Funding: The project was financially supported by the National Natural Science Foundation of China (31801320 and 31761143004), the Department of Science and Technology of Gansu Province (20JR10RA545), and the Fostering Foundation for Excellent Ph.D. Dissertations of Gansu Agricultural University (YB2018002).

Institutional Review Board Statement: Not applicable.

Informed Consent Statement: Not applicable.

Data Availability Statement: All data and material produced in the present study are provided in this manuscript and the supplementary materials. All high-throughput sequencing data were deposited in the NCBI Sequence Read Archive (SRA) database under accession number PRJNA689959.

Acknowledgments: We appreciate the excellent technical assistance for the field work and laboratory tests provided by undergraduate and graduate students at the Gansu Agricultural University Rainfed Agricultural Experimental Station.

Conflicts of Interest: The authors declare no conflict of interest.

References

1. Fierer, N. Embracing the unknown: Disentangling the complexities of the soil microbiome. *Nat. Rev. Microbiol.* **2017**, *15*, 579–590. [CrossRef] [PubMed]
2. Berendsen, R.L.; Pieterse, C.M.; Bakker, P.A. The rhizosphere microbiome and plant health. *Trends Plant Sci.* **2012**, *17*, 478–486. [CrossRef]
3. Nannipieri, P.; Ascher, J.; Ceccherini, M.; Landi, L.; Pietramellara, G.; Renella, G. Microbial diversity and soil functions. *Eur. J. Soil Sci.* **2003**, *54*, 655–670. [CrossRef]
4. Miltner, A.; Richnow, H.-H.; Kopinke, F.-D.; Kästner, M. Assimilation of CO_2 by soil microorganisms and transformation into soil organic matter. *Org. Geochem.* **2004**, *35*, 1015–1024. [CrossRef]
5. Lacis, A.A.; Schmidt, G.A.; Rind, D.; Ruedy, R.A. Atmospheric CO_2: Principal control knob governing Earth's temperature. *Science* **2010**, *330*, 356–359. [CrossRef] [PubMed]
6. Tolli, J.; King, G.M. Diversity and structure of bacterial chemolithotrophic communities in pine forest and agroecosystem soils. *Appl. Environ. Microbiol.* **2005**, *71*, 8411–8418. [CrossRef]
7. Stein, S.; Selesi, D.; Schilling, R.; Pattis, I.; Schmid, M.; Hartmann, A. Microbial activity and bacterial composition of H_2-treated soils with net CO_2 fixation. *Soil Biol. Biochem.* **2005**, *37*, 1938–1945. [CrossRef]
8. Ensign, S.A.; Small, F.J.; Allen, J.R.; Sluis, M.K. New roles for CO_2 in the microbial metabolism of aliphatic epoxides and ketones. *Arch. Microbiol.* **1998**, *169*, 179–187. [CrossRef]
9. Cai, P.; Ning, Z.; Zhang, M.; Guo, C.; Niu, M.; Shi, J. Autotrophic metabolism considered to extend the applicability of the carbon balances model for assessing biodegradation in petroleum-hydrocarbon-contaminated aquifers with abnormally low dissolved inorganic carbon. *J. Clean. Prod.* **2020**, *261*, 120738. [CrossRef]
10. Tabita, F.R. Microbial ribulose 1,5-bisphosphate carboxylase/oxygenase: A different perspective. *Photosynth. Res.* **1999**, *60*, 1–28. [CrossRef]
11. Wildman, S.G. Along the trail from Fraction I protein to Rubisco (ribulose bisphosphate carboxylase-oxygenase). *Photosynth. Res.* **2002**, *73*, 243–250. [CrossRef] [PubMed]
12. Correa, S.S.; Schultz, J.; Lauersen, K.J.; Rosado, A.S. Natural carbon fixation and advances in synthetic engineering for redesigning and creating new fixation pathways. *J. Adv. Res.* **2022**, *in press*. [CrossRef]
13. Tabita, F.R. Molecular and cellular regulation of autotrophic carbon dioxide fixation in microorganisms. *Microbiol. Rev.* **1988**, *52*, 155–189. [CrossRef] [PubMed]
14. Selesi, D.; Pattis, I.; Schmid, M.; Kandeler, E.; Hartmann, A. Quantification of bacterial RubisCO genes in soils by *cbbL* targeted real-time PCR. *J. Microbiol. Methods* **2007**, *69*, 497–503. [CrossRef] [PubMed]
15. Kusian, B.; Bowien, B. Organization and regulation of cbb CO_2 assimilation genes in autotrophic bacteria. *FEMS Microb. Ecol.* **1997**, *21*, 135–155. [CrossRef]
16. Selesi, D.; Schmid, M.; Hartmann, A. Diversity of green-like and red-like ribulose-1,5-bisphosphate carboxylase/oxygenase large-subunit genes (*cbbL*) in differently managed agricultural soils. *Appl. Environ. Microbiol.* **2005**, *71*, 175–184. [CrossRef] [PubMed]
17. Navarro-Noya, Y.E.; Gómez-Acata, S.; Montoya-Ciriaco, N.; Rojas-Valdez, A.; Suárez-Arriaga, M.C.; Valenzuela-Encinas, C.; Jiménez-Bueno, N.; Verhulst, N.; Govaerts, B.; Dendooven, L. Relative impacts of tillage, residue management and crop-rotation on soil bacterial communities in a semi-arid agroecosystem. *Soil Biol. Biochem.* **2013**, *65*, 86–95. [CrossRef]
18. Zhang, B.; Li, Y.; Ren, T.; Tian, Z.; Wang, G.; He, X.; Tian, C. Short-term effect of tillage and crop rotation on microbial community structure and enzyme activities of a clay loam soil. *Biol. Fertil. Soils* **2014**, *50*, 1077–1085. [CrossRef]
19. Lamptey, S.; Li, L.; Xie, J.; Coulter, J.A. Tillage system affects soil water and photosynthesis of plastic-mulched maize on the semiarid Loess Plateau of China. *Soil Tillage Res.* **2020**, *196*, 104479. [CrossRef]

20. Klik, A.; Rosner, J. Long-term experience with conservation tillage practices in Austria: Impacts on soil erosion processes. *Soil Tillage Res.* **2020**, *203*, 104669. [CrossRef]
21. Scopel, E.; Triomphe, B.; Affholder, F.; Da Silva, F.A.M.; Corbeels, M.; Xavier, J.H.V.; Lahmar, R.; Recous, S.; Bernoux, M.; Blanchart, E. Conservation agriculture cropping systems in temperate and tropical conditions, performances and impacts. A review. *Agron. Sustain. Dev.* **2013**, *33*, 113–130. [CrossRef]
22. Aziz, I.; Mahmood, T.; Islam, K.R. Effect of long term no-till and conventional tillage practices on soil quality. *Soil Tillage Res.* **2013**, *131*, 28–35. [CrossRef]
23. Liu, C.; Cutforth, H.; Chai, Q.; Gan, Y. Farming tactics to reduce the carbon footprint of crop cultivation in semiarid areas. A review. *Agron. Sustain. Dev.* **2016**, *36*, 69. [CrossRef]
24. Sarker, J.R.; Singh, B.P.; Cowie, A.L.; Fang, Y.; Collins, D.; Badgery, W.; Dalal, R.C. Agricultural management practices impacted carbon and nutrient concentrations in soil aggregates, with minimal influence on aggregate stability and total carbon and nutrient stocks in contrasting soils. *Soil Tillage Res.* **2018**, *178*, 209–223. [CrossRef]
25. Martínez-Mena, M.; Perez, M.; Almagro, M.; Garcia-Franco, N.; Díaz-Pereira, E. Long-term effects of sustainable management practices on soil properties and crop yields in rainfed Mediterranean almond agroecosystems. *Eur. J. Agron.* **2021**, *123*, 126207. [CrossRef]
26. Chai, Q.; Nemecek, T.; Liang, C.; Zhao, C.; Yu, A.; Coulter, J.A.; Wang, Y.; Hu, F.; Wang, L.; Siddique, K.H.M.; et al. Integrated farming with intercropping increases food production while reducing environmental footprint. *Proc. Natl. Acad. Sci. USA* **2021**, *118*, e2106382118. [CrossRef] [PubMed]
27. Buffett, H.G. Reaping the benefits of no-tillage farming. *Nature* **2012**, *484*, 455. [CrossRef] [PubMed]
28. Liu, K.; Bandara, M.; Hamel, C.; Knight, J.D.; Gan, Y. Intensifying crop rotations with pulse crops enhances system productivity and soil organic carbon in semi-arid environments. *Field Crops Res.* **2020**, *248*, 107657. [CrossRef]
29. Wang, L.; Gan, Y.; Bainard, L.; Hamel, C.; St-Arnaud, M.; Hijri, M. Expression of N-cycling genes of root microbiomes provides insights for sustaining oilseed crop production. *Environ. Microbiol.* **2020**, *22*, 4545–4556. [CrossRef]
30. Borrell, A.N.; Shi, Y.; Gan, Y.; Bainard, L.D.; Germida, J.J.; Hamel, C. Fungal diversity associated with pulses and its influence on the subsequent wheat crop in the Canadian prairies. *Plant Soil* **2017**, *414*, 13–31. [CrossRef]
31. Liu, C.; Li, L.; Xie, J.; Coulter, J.A.; Zhang, R.; Luo, Z.; Cai, L.; Wang, L.; Gopalakrishnan, S. Soil bacterial diversity and potential functions are regulated by long-term conservation tillage and straw mulching. *Microorganisms* **2020**, *8*, 836. [CrossRef]
32. Ding, Y.; Wang, F.; Mu, Q.; Sun, Y.; Cai, H.; Zhou, Z.; Xu, J.; Shi, H. Estimating land use/land cover change impacts on vegetation response to drought under 'Grain for Green' in the Loess Plateau. *Land Degrad. Dev.* **2021**, *32*, 5083–5098. [CrossRef]
33. Jiang, F.; Xie, X.; Liang, S.; Wang, Y.; Zhu, B.; Zhang, X.; Chen, Y. Loess Plateau evapotranspiration intensified by land surface radiative forcing associated with ecological restoration. *Agric. For. Meteorol.* **2021**, *311*, 108669. [CrossRef]
34. Luo, Z.; Gan, Y.; Niu, Y.; Zhang, R.; Li, L.; Cai, L.; Xie, J. Soil quality indicators and crop yield under long-term tillage systems. *Exp. Agric.* **2017**, *53*, 497–511. [CrossRef]
35. Li, Y.; Gan, Y.; Lupwayi, N.; Hamel, C. Influence of introduced arbuscular mycorrhizal fungi and phosphorus sources on plant traits, soil properties, and rhizosphere microbial communities in organic legume-flax rotation. *Plant Soil* **2019**, *443*, 87–106. [CrossRef]
36. Li, Y.; Laterrière, M.; Lay, C.-Y.; Klabi, R.; Masse, J.; St-Arnaud, M.; Yergeau, É.; Lupwayi, N.Z.; Gan, Y.; Hamel, C. Effects of arbuscular mycorrhizal fungi inoculation and crop sequence on root-associated microbiome, crop productivity and nutrient uptake in wheat-based and flax-based cropping systems. *Appl. Soil Ecol.* **2021**, *168*, 104136. [CrossRef]
37. Hossain, Z.; Hubbard, M.; Gan, Y.; Bainard, L.D. Root rot alters the root-associated microbiome of field pea in commercial crop production systems. *Plant Soil* **2021**, *460*, 593–607. [CrossRef]
38. Juliet, P.M.; Lynne, B.; Randerson, P.F. Analysis of microbial community functional diversity using sole-carbon-source utilisation profiles—A critique. *FEMS Microbiol. Ecol.* **2002**, *42*, 1–14. [CrossRef]
39. Garland, J.L.; Mills, A.L. Classification and characterization of heterotrophic microbial communities on the basis of patterns of community-level sole-carbon-source utilization. *Appl. Environ. Microbiol.* **1991**, *57*, 2351–2359. [CrossRef]
40. Li, P.; Chen, W.; Han, Y.; Wang, D.; Zhang, Y.; Wu, C. Effects of straw and its biochar applications on the abundance and community structure of CO_2-fixing bacteria in a sandy agricultural soil. *J. Soils Sed.* **2020**, *20*, 2225–2235. [CrossRef]
41. Li, Z.; Tong, D.; Nie, X.; Xiao, H.; Jiao, P.; Jiang, J.; Li, Q.; Liao, W. New insight into soil carbon fixation rate: The intensive co-occurrence network of autotrophic bacteria increases the carbon fixation rate in depositional sites. *Agric. Ecosyst. Environ.* **2021**, *320*, 107579. [CrossRef]
42. Li, M.; He, P.; Guo, X.-L.; Zhang, X.; Li, L.-J. Fifteen-year no tillage of a Mollisol with residue retention indirectly affects topsoil bacterial community by altering soil properties. *Soil Tillage Res.* **2021**, *205*, 104804. [CrossRef]
43. Lynn, T.M.; Ge, T.; Yuan, H.; Wei, X.; Wu, X.; Xiao, K.; Kumaresan, D.; Yu, S.S.; Wu, J.; Whiteley, A.S. Soil carbon-fixation rates and associated bacterial diversity and abundance in three natural ecosystems. *Microb. Ecol.* **2017**, *73*, 645–657. [CrossRef]
44. Zuber, S.M.; Villamil, M.B. Meta-analysis approach to assess effect of tillage on microbial biomass and enzyme activities. *Soil Biol. Biochem.* **2016**, *97*, 176–187. [CrossRef]
45. Sun, B.; Jia, S.; Zhang, S.; Mclaughlin, N.B.; Liang, A.; Chen, X.; Liu, S.; Zhang, X. No tillage combined with crop rotation improves soil microbial community composition and metabolic activity. *Environ. Sci. Pollut. Res.* **2016**, *23*, 6472–6482. [CrossRef]

46. Fiorini, A.; Boselli, R.; Maris, S.C.; Santelli, S.; Ardenti, F.; Capra, F.; Tabaglio, V. May conservation tillage enhance soil C and N accumulation without decreasing yield in intensive irrigated croplands? Results from an eight-year maize monoculture. *Agric. Ecosyst. Environ.* **2020**, *296*, 106926. [CrossRef]

47. Wei, H.; Peng, C.; Yang, B.; Song, H.; Li, Q.; Jiang, L.; Wei, G.; Wang, K.; Wang, H.; Liu, S. Contrasting soil bacterial community, diversity, and function in two forests in China. *Front. Microbiol.* **2018**, *9*, 1693. [CrossRef] [PubMed]

48. Deng, J.; Zhang, Y.; Yin, Y.; Zhu, X.; Zhu, W.; Zhou, Y.J.P. Comparison of soil bacterial community and functional characteristics following afforestation in the semi-arid areas. *PeerJ* **2019**, *7*, e7141. [CrossRef]

49. Kaiser, K.; Wemheuer, B.; Korolkow, V.; Wemheuer, F.; Nacke, H.; Schöning, I.; Schrumpf, M.; Daniel, R. Driving forces of soil bacterial community structure, diversity, and function in temperate grasslands and forests. *Sci. Rep.* **2016**, *6*, 33696. [CrossRef]

50. Ren, B.; Hu, Y.; Chen, B.; Zhang, Y.; Thiele, J.; Shi, R.; Liu, M.; Bu, R. Soil pH and plant diversity shape soil bacterial community structure in the active layer across the latitudinal gradients in continuous permafrost region of Northeastern China. *Sci. Rep.* **2018**, *8*, 5619. [CrossRef]

51. Qin, J.; Li, M.; Zhang, H.; Liu, H.; Zhao, J.; Yang, D. Nitrogen deposition reduces the diversity and abundance of *cbbL* gene-containing CO_2-fixing microorganisms in the soil of the stipa baicalensis steppe. *Front. Microbiol.* **2021**, *12*, 575908. [CrossRef] [PubMed]

52. Geisseler, D.; Scow, K.M. Long-term effects of mineral fertilizers on soil microorganisms—A review. *Soil Biol. Biochem.* **2014**, *75*, 54–63. [CrossRef]

53. Niu, Y.; Bainard, L.D.; May, W.E.; Hossain, Z.; Hamel, C.; Gan, Y. Intensified pulse rotations buildup pea rhizosphere pathogens in cereal and pulse based cropping systems. *Front. Microbiol.* **2018**, *9*, 1909. [CrossRef] [PubMed]

54. Hamel, C.; Gan, Y.; Sokolski, S.; Bainard, L.D. High frequency cropping of pulses modifies soil nitrogen level and the rhizosphere bacterial microbiome in 4-year rotation systems of the semiarid prairie. *Appl. Soil Ecol.* **2018**, *126*, 47–56. [CrossRef]

55. Nannipieri, P.; Ascher-Jenull, J.; Ceccherini, M.T.; Pietramellara, G.; Renella, G.; Schloter, M. Beyond microbial diversity for predicting soil functions: A mini review. *Pedosphere* **2020**, *30*, 5–17. [CrossRef]

56. Inubushi, K.; Yashima, M.; Hanazawa, S.; Goto, A.; Miyamoto, K.; Tsuboi, T.; Asea, G. Long-term fertilizer management in NERICA cultivated upland affects on soil bio-chemical properties. *Soil Sci. Plant Nutr.* **2020**, *66*, 247–253. [CrossRef]

57. Wang, X.; Teng, Y.; Ren, W.; Han, Y.; Wang, X.; Li, X. Soil bacterial diversity and functionality are driven by plant species for enhancing polycyclic aromatic hydrocarbons dissipation in soils. *Sci. Total Environ.* **2021**, *797*, 149204. [CrossRef]

58. Tang, C.; Li, Y.; Song, J.; Antonietti, M.; Yang, F. Artificial humic substances improve microbial activity for binding CO_2. *iScience* **2021**, *24*, 102647. [CrossRef]

59. Wang, Q.; Liang, A.; Chen, X.; Zhang, S.; Zhang, Y.; McLaughlin, N.B.; Gao, Y.; Jia, S. The impact of cropping system, tillage and season on shaping soil fungal community in a long-term field trial. *Eur. J. Soil Biol.* **2021**, *102*, 103253. [CrossRef]

60. Lu, J.; Qiu, K.; Li, W.; Wu, Y.; Ti, J.; Chen, F.; Wen, X. Tillage systems influence the abundance and composition of autotrophic CO_2-fixing bacteria in wheat soils in North China. *Eur. J. Soil Biol.* **2019**, *93*, 103086. [CrossRef]

61. Essel, E.; Xie, J.; Deng, C.; Peng, Z.; Wang, J.; Shen, J.; Xie, J.; Coulter, J.A.; Li, L. Bacterial and fungal diversity in rhizosphere and bulk soil under different long-term tillage and cereal/legume rotation. *Soil Tillage Res.* **2019**, *194*, 104302. [CrossRef]

62. St. Luce, M.; Lemke, R.; Gan, Y.; McConkey, B.; May, W.; Campbell, C.; Zentner, R.; Wang, H.; Kroebel, R.; Fernandez, M.; et al. Diversifying cropping systems enhances productivity, stability, and nitrogen use efficiency. *Agron. J.* **2020**, *112*, 1517–1536. [CrossRef]

agriculture

Article

Spectroscopic Investigation on the Effects of Biochar and Soluble Phosphorus on Grass Clipping Vermicomposting

Etelvino Henrique Novotny [1,*], Fabiano de Carvalho Balieiro [1], Ruben Auccaise [2], Vinícius de Melo Benites [1] and Heitor Luiz da Costa Coutinho [1,†]

1 Embrapa Solos, Rua Jardim Botânico, 1024, Rio de Janeiro 22460-000, Brazil; fabiano.balieiro@embrapa.br (F.d.C.B.); vinicius.benites@embrapa.br (V.d.M.B.)
2 Departamento de Física, Universidade Estadual de Ponta Grossa, Av. General Carlos Cavalcanti 4748, Ponta Grossa 84030-900, Brazil; raestrada@uepg.br
* Correspondence: etelvino.novotny@embrapa.br
† In memorian.

Abstract: Seeking to evaluate the hypothesis that biochar optimises the composting and vermicomposting processes as well as their product quality, we carried out field and greenhouse experiments. Four grass clipping composting treatments (only grass, grass + single superphosphate (SSP), grass + biochar and grass + SSP + biochar) were evaluated. At the end of the maturation period (150 days), the composts were submitted to vermicomposting (Eisenia fetida earthworm) for an additional 90 days. Ordinary fine charcoal was selected due to its low cost (a by-product of charcoal production) and great availability; this is important since the obtained product presents low commercial value. A greater maturity of the organic matter (humification) was observed in the vermicompost treatments compared with the compost-only treatments. The addition of phosphate significantly reduced the pH (from 6.7 to 4.8), doubled the electrical conductivity and inhibited biological activity, resulting in less than 2% of the number of earthworms found in the treatment without phosphate. The addition of soluble phosphate inhibited the humification process, resulting in a less-stable compound with the preservation of labile structures, primarily cellulose. The P species found corroborate these findings because the pyrophosphate conversion from SSP in the absence of biochar may explain the strong acidification and increased electric conductivity. Biochar appears to prevent this conversion, thus mitigating the deleterious effects of SSP and favouring the formation of organic P species from SSP (78.5% of P in organic form with biochar compared to only 12.8% in the treatments without biochar). In short, biochar decreases pyrophosphate formation from SSP, avoiding acidification and salinity; therefore, biochar improves the whole composting and vermicomposting process and product quality. Vermicompost with SSP and biochar should be tested as a soil conditioner on account of its greater proportion of stabilized C and organic P.

Keywords: ^{13}C nuclear magnetic resonance; ^{31}P nuclear magnetic resonance; charcoal; *Eisenia fetida*; pyrogenic carbon

Citation: Novotny, E.H.; Balieiro, F.d.C.; Auccaise, R.; Benites, V.d.M.; Coutinho, H.L.d.C. Spectroscopic Investigation on the Effects of Biochar and Soluble Phosphorus on Grass Clipping Vermicomposting. *Agriculture* **2022**, *12*, 1011. https://doi.org/10.3390/agriculture12071011

Academic Editor: Borbála Biró

Received: 25 April 2022
Accepted: 7 July 2022
Published: 13 July 2022

Publisher's Note: MDPI stays neutral with regard to jurisdictional claims in published maps and institutional affiliations.

1. Introduction

Biochar is a C-rich product distinct from charcoal because it is produced for soil application purposes in order to improve its quality (organic matter stability, cation exchange capacity (CEC), soil fertility, water retention capacity, porosity and biological activity are commonly increased), prevent nutrient leaching, improve C storage or purify the soil from pollutants [1–3]. The uses of and studies regarding this pyrolyzed biomass for agriculture have increased in recent decades, in part due to recognition of the role of carbonised biomass in the high fertility of *terra preta de indios* soils, which are anthropic soils that had a high input of carbonised biomass during the pre-Colombian period [4–6]. Although a number of studies have investigated the effects of different pyrolyzed residues on plant productivity and chemical and physical soil properties [7–11], little is known about *biochar* interactions with soil microbiota and fauna or with the native microorganisms from the raw materials used in composting [8,12–14].

Composting is a technique in which microbial activity during organic material decay is optimised and the material is transformed into either a fertiliser or soil conditioner. Vermicomposting, on the other hand, is a process by which earthworms aid in recycling and increase the velocity of organic matter stabilisation through residue homogenisation. The degradation of cellulose and inoculation of other organisms are stimulated by these macrofauna representatives, and humification tends to increase in the presence of these organisms [15–17].

Although earthworms occur in most environments, these organisms are sensitive to soil management practices [18,19], trace metals (mainly Cd) [20,21] and carbon polyaromatic-containing substances [21,22]; thus, they can be used as quality indicators for soil or different organic substrates [21,23,24]. Fertilisers can affect the abundance and density of earthworms in the soil, although limited studies have investigated the effect of soluble sources, including phosphorus (P), on these soil organisms. Sarathchandra et al. [25] performed field studies and did not observe changes in the populations of *Lumbricus rubellus* or *Aporrectodea caliginosa* in either the adult or young forms after three years of single superphosphate (SSP) and rock phosphate application in North Carolina. However, laboratory studies have shown that direct contact with SSP can be lethal to the Californian red earthworm (*Eisenia fetida*), which has led to studies recommending the monitoring of these animal populations after applications of inorganic and soluble sources of P to the field [26].

In order to find a way to minimise the external input (fertilisers and soil conditioner) needed to maintain the extensive grasslands at the Rio de Janeiro International Airport in Brazil, which produces approximately 2 Mg of grass clippings daily, and also to reduce the disposal cost of this low-density raw material and hence reduce the environmental impact of the airport, the potential use of grass clippings for compost was proposed [27,28]. Therefore, this study aimed to evaluate the impact of using soluble phosphate and biochar on the establishment and development of the Californian red earthworm *E. fetida* in compost and vermicompost that included grass clippings from the Rio de Janeiro International Airport as well as evaluate the final products using ^{13}C and ^{31}P solid-state nuclear magnetic resonance spectroscopy. We hypothesized that the introduction of P in the composting process would increase the nutritional status of the compost/vermicompost, since P is a limiting element to plant growth in tropical oxidized soils, and that biochar would improve the composting and vermicomposting processes as well as the soil conditions for grass growth in the airport field area (not tested).

2. Materials and Methods

2.1. Composting Site and Origin of Raw Material

The experiment was conducted in a covered hangar of the Rio de Janeiro/Galeão International Airport ($-22.804114°$; $-43.229841°$), which is located in Rio de Janeiro, Brazil. Composting was performed using grass clippings obtained by mowing the airport's islands, i.e., the unused grassy areas between taxiways, between runways or between a taxiway and a runway. The clippings were ground to homogenise the particle size. The chemical composition of the residue (mean $\pm$ standard deviation, n=20) was as follows [28]: C = 415 $\pm$ 2 g kg^{-1}; N = 11 $\pm$ 2 g kg^{-1}; K = 10.7 $\pm$ 0.7 g kg^{-1}; Ca = 4.2 $\pm$ 0.7 g kg^{-1}; Mg = 1.3 $\pm$ 0.1 g kg^{-1}; P = 0.5 $\pm$ 0.1 g kg^{-1}; and C/N ratio = 47 $\pm$ 7.5. At the beginning of the experiment, the average moisture and density of the grass clippings were 70–80% and ~80 kg m^{-3}, respectively.

2.2. Composting of Grass Clippings

Four treatments arranged in a full factorial design (2^2: two levels of phosphate and of *biochar* in grass clipping composting) were tested: grass (G); grass + biochar (GB); grass + phosphate (GP); and grass + biochar + phosphate (GBP). The phosphate used was single superphosphate (SSP) with 8.7% P, mainly in the form of monocalcium phosphate (). Previous composting was performed in 2.5 m^3 piles. The charcoal used in the piles was the residue of vegetable charcoal production (*Eucalyptus sp.*, conventional carbonisation at 450 °C for six hours). This residue was a fine-grained charcoal, with a grain size usually smaller than 5 mm. The used sample exhibited non-uniform and small grain sizes, with only 10% of the mass retained on a 4-mm sieve. The quantities of each input used to prepare the compost piles are presented in Table 1.

Table 1. Quantities (kg) of each material/input used in the composting piles (2.5 m^3).

Treatments	Grass (kg)	Biochar (kg)	Phosphate (kg)
Grass	200	-	-
Grass + biochar	160	143	-
Grass + phosphate	200	-	25.6
Grass + phosphate + biochar	160	143	25.6

At the beginning of the process, the piles were manually turned every two days. After 30 days, the piles were turned every 10 days. After 120 days of composting, all of the piles were ground to 5 cm, and the piles were turned twice a week for 30 days.

After 150 days of composting, six sub-samples of each pile were collected and transferred to plastic boxes for vermicomposting.

2.3. Vermicomposting

Vermicomposting was performed for 90 days using a completely randomised experimental design with six replicates. Twenty adult *E. fetida* were added to 10 L of the different composts and placed in 20 L plastic boxes. *E. fetida* earthworms were chosen because they exhibit high prolificacy, precociousness, survival rates and adaptability to captivity [4]. The experimental design was a 2 × 4 full factorial design (with and without earthworms × 4 former treatments, see Section 2.2) with 6 replicates and a total of 48 worm boxes.

Compost moisture was kept constant with frequent wettings. After 90 days, the earthworms from each box were counted.

2.4. Physical, Chemical and Spectroscopy Characterisation

Sub-samples from each box were dried at 45 °C for 48 h and ground, and the C and N concentrations were determined using a PerkinElmer 2400 CHNS (Waltham, MA, USA) elemental analyser. The pH and electrical conductivity (EC) of the different composts were determined as follows: approximately 15 g of compost was weighed in lidded plastic containers and then 85 mL of distilled water was added. The samples were stirred using a horizontal agitator for 30 min at 40 rpm and left to stand for 20 min, after which the EC and pH were measured.

For spectroscopic characterisation, sub-samples of each treatment were freeze-dried and ground in liquid N_2. Solid-state ^{13}C and ^{31}P nuclear magnetic resonance (NMR) spectra were obtained with a Varian Inova (11.74 T) spectrometer at ^{13}C, ^{31}P and ^{1}H frequencies of 125.7, 202.5 and 500.0 MHz, respectively. For this, a T3NB HXY with a 4-mm probe was utilised to detect the ^{13}C and ^{31}P nuclei and the rotors were spun using dry air at 15 kHz. All experiments were carried out at room temperature.

Two NMR pulse procedures were applied: variable amplitude cross-polarisation (CP) for ^{13}C and direct polarisation (DP) for ^{31}P.

In the ^{13}C CP-MAS experiment an optimised recycle delay (d1) of 500 ms was used; the ^{1}H 90-pulse was set to 3 μs, the contact time value to 1 ms and the acquisition time to 15 ms. High-power two-pulse phase modulation (TPPM) ^{1}H decoupling at 70 kHz was employed. The cross-polarisation time was selected after variable contact time experiments and the recycle delays were selected to be five times longer than the longest spin–lattice relaxation time (T_1), as determined by inversion recovery experiments.

In the ^{31}P DP-MAS experiment, the recycle delay was 10 min (after inversion-recovery experiments), the ^{31}P 90-pulse was set to 3 µs and the acquisition time to 15 ms.

2.5. Data Analysis

The data from the vermicomposting experiment (pH, EC and number of earthworms) were subjected to a multivariate analysis of variance (MANOVA), while the data from composting and further vermicomposting were analysed using repeated measures MANOVA. The normality and homoscedasticity of the residuals were tested and, when necessary, appropriate data transformation (log for C and C/N ratio and square root for the number of earthworms) was applied. When a statistically significant effect was detected, the means were compared using Duncan's test at $p < 0.05$. The software Statistica 7.1 [29] was used for all statistical analyses.

The multivariate analysis of the spectral data was performed using the software Unscrambler 10.4 (CAMO, Norway). Principal component analysis (PCA) was carried out using the full ^{13}C NMR spectra obtained, after area normalisation and mean-centring of the data. To aid in the analyses of the ^{31}P NMR results, the multivariate curve resolution (MCR) procedure was carried out. The basic goals of MCR are: the determination of the number of components co-existing in the chemical system; the extraction of the pure spectra of the components (qualitative analysis); and the extraction of the concentration profiles of these components (quantitative analysis). This analysis was preceded by PCA to estimate the number of components in the mixture. After this, the rotation of the PC was calculated without orthogonality constraints (in this way, it would have infinite solutions). To solve this, new constraints were adopted (e.g., non-negative concentrations and/or non-negative spectra). In this way, when the goals of MCR are achieved, it is possible to unravel the "true" underlying sources of data variation, after which the results with physical meaning are easily interpretable [18].

3. Results

3.1. Earthworm Reproduction with Different Substrates

The application of only phosphate drastically inhibited earthworm reproduction (Figure 1a). However, the interaction between the main factors of phosphate and biochar was statistically significant ($p = 2.6 \times 10^{-6}$), with this detrimental effect mitigated by biochar, resulting in significantly larger populations than the GP treatment but still lower populations than the treatments without phosphate (G and GB).

3.2. Chemical and Spectroscopy Analysis of the Composts and Vermicomposts

The addition of phosphate fertiliser increased the medium's acidity and the EC of the vermicompost; however, biochar decreased these effects significantly (Figure 1b,c). Biochar prevented the pronounced acidification caused by phosphate; however, the pH was still lower than that of the treatments without phosphate (significant interaction, $p = 2.0 \times 10^{-6}$). Meanwhile, for EC just the main factors (phosphate and biochar) were statistically significant ($p = 5.5 \times 10^{-6}$ and 9.9×10^{-4}, respectively), without significant interaction; therefore, the biochar decreased the EC, fully mitigating the deleterious effect of phosphate application on the vermicompost EC.

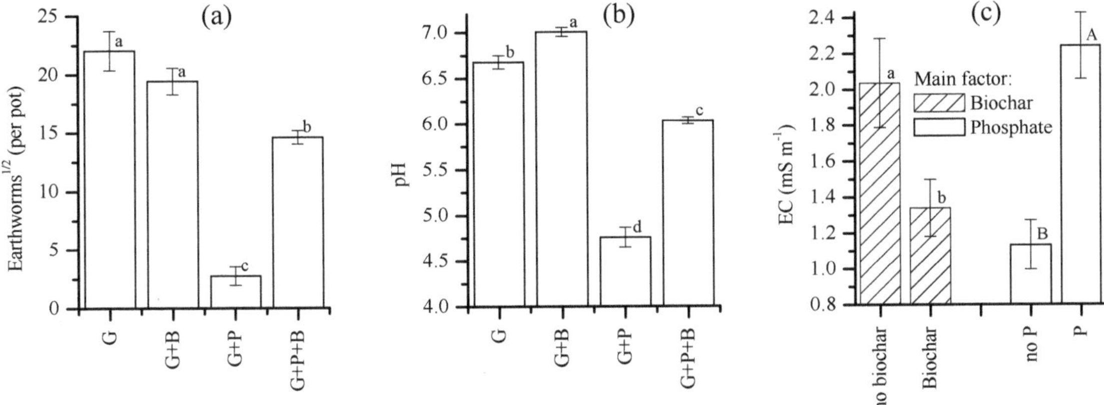

Figure 1. Square root of earthworm number after inoculation for 90 days in different composted substrates (**a**); pH (**b**) and electrical conductivity (EC) (**c**) of the vermicompost. G: grass clippings; G + B: grass + biochar; G + P: grass + phosphate; and G + B + P: grass + biochar + phosphate. Columns with the same lower-case letters (for earthworms, pH and biochar factor for EC) and upper-case letters (for P factor for EC) indicate no statistical difference at $p < 0.05$ using Duncan test. Vertical bars denote standard errors.

The introduction of biochar increased C concentrations in all substrates (Figure 2a), and no significant effect was observed for phosphate or vermicomposting. Concerning N concentration and C/N ratio, the interaction of phosphate × biochar was significant ($p = 2.8 \times 10^{-4}$ and 2.5×10^{-4}, respectively), being that phosphate and biochar decreased N and increased the C/N ratio (Figure 2b), with the strongest effect being from biochar. In addition, further vermicomposting increased the N content and decreased the C/N ratio (Figure 2c), indicating further humification.

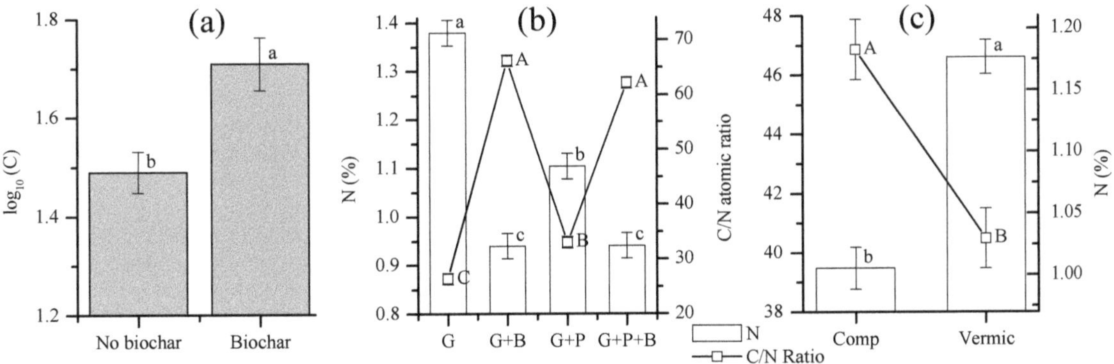

Figure 2. Biochar effect on C concentration (**a**), phosphate and biochar interaction effects on N concentration and on C/N ratio (**b**), and effects of vermicomposting on these variables (**c**). G: grass clippings; G + B: grass + biochar; G + P: grass + phosphate; and G + B + P: grass + biochar + phosphate; Comp: composting; and Vermic: further vermicomposting. Columns with the same lower-case letters (C and N) and symbols with the same upper-case letters (C/N ratio) indicate no statistical difference at $p < 0.05$ using Duncan test. Vertical bars denote standard errors.

The change in humification was confirmed by the ^{13}C and ^{31}P NMR spectra, which are shown below. The C/N ratio increased in all treatments with biochar (Figure 2b), which was also explained by the inclusion of material rich in recalcitrant C and poor in N. Meanwhile, a significant effect of vermicomposting we observed was that it decreased the C/N ratio. The addition of phosphate to the substrates in the absence of biochar resulted in higher C/N ratios in the compost and vermicompost (no interaction), confirming the inhibition of humification by phosphate.

The organic chemical structures detected using ^{13}C NMR (Figure 3) indicate that, following composting, the material still exhibited significant amounts of labile structures, mainly cellulose (O-alkyl, with signals at ~65 and 73 ppm and di-O-alkyl, with signal at ~105 ppm). The relative contribution of these regions varied between substrates, indicating selective degradation of cellulose (higher biological activity and higher humification) in certain substrates. For the substrates with biochar, a pronounced increase of the C-aryl signal was observed (~125 ppm), which is a typical signal of condensed aromatic rings. These differences were summarised and highlighted using PCA and are discussed in detail below.

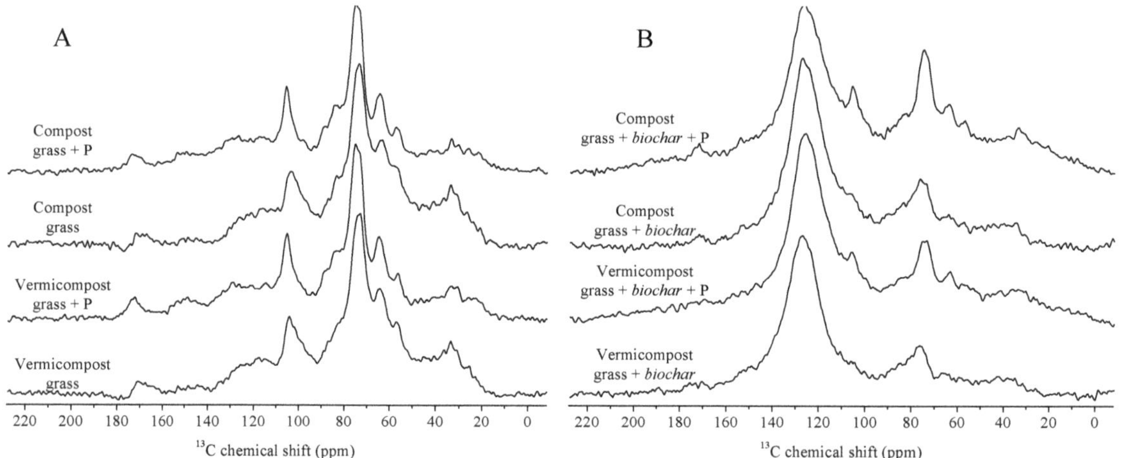

Figure 3. Solid state ^{13}C NMR spectra of the different compost and vermicompost obtained. Without biochar (**A**) and with biochar (**B**).

3.3. Progress of Humification According to Principal Component Analysis (PCA)

Using PCA, we were able to satisfactorily model the data (97% of variance captured) with two principal components (PCs). The first PC (92% of total variance) identified the different substrates exhibiting bipolar loadings (Figure 4A), with positive loadings for the sp^2 C signals of biochar (aryl) and negative for the sp^3 hybridised C of grass cellulose (O-alkyl and di-O-alkyl). The second PC (5% of total variance) was characterised by negative loadings for the signals typical of labile structures (Figure 4A), primarily cellulose, that were partially oxidised to uronic acids (O-alkyl, di-O-alkyl and carboxyl with a signal at ~173 ppm). Therefore, this PC served as an indicator of the progress of humification, since less labile structures indicate the progression of the humification process.

The samples with biochar (GB, VGB, GBP, VGBP) presented the highest scores for PC1 (Figure 4B), confirming the contribution of polycondensed aryl structures towards their ^{13}C NMR spectra.

The samples containing only grass, with and without earthworms, exhibited the highest PC2 scores, indicating a lower proportion of labile structures and more advanced humification (Figure 4B). Treatments with phosphate and without biochar (compost—GP and vermicompost—VGP) exhibited the lowest scores for PC2 (Figure 4B), indicating that phosphate inhibited the advance of humification by suppressing the biological activity of micro- and macro-organisms (after the earthworm population decrease, Figure 1a), resulting in a material rich in labile structures, i.e., partially oxidised cellulose. This finding confirmed the results for the N concentrations and C/N ratios as well as the detrimental effects on earthworms. The samples with biochar (GB, VGB, GBP, VGBP) exhibited intermediate scores (Figure 4B) for this component (uronic acids), indicating that the maturation of the compost and vermicompost were similar for the samples with biochar, regardless of phosphate presence.

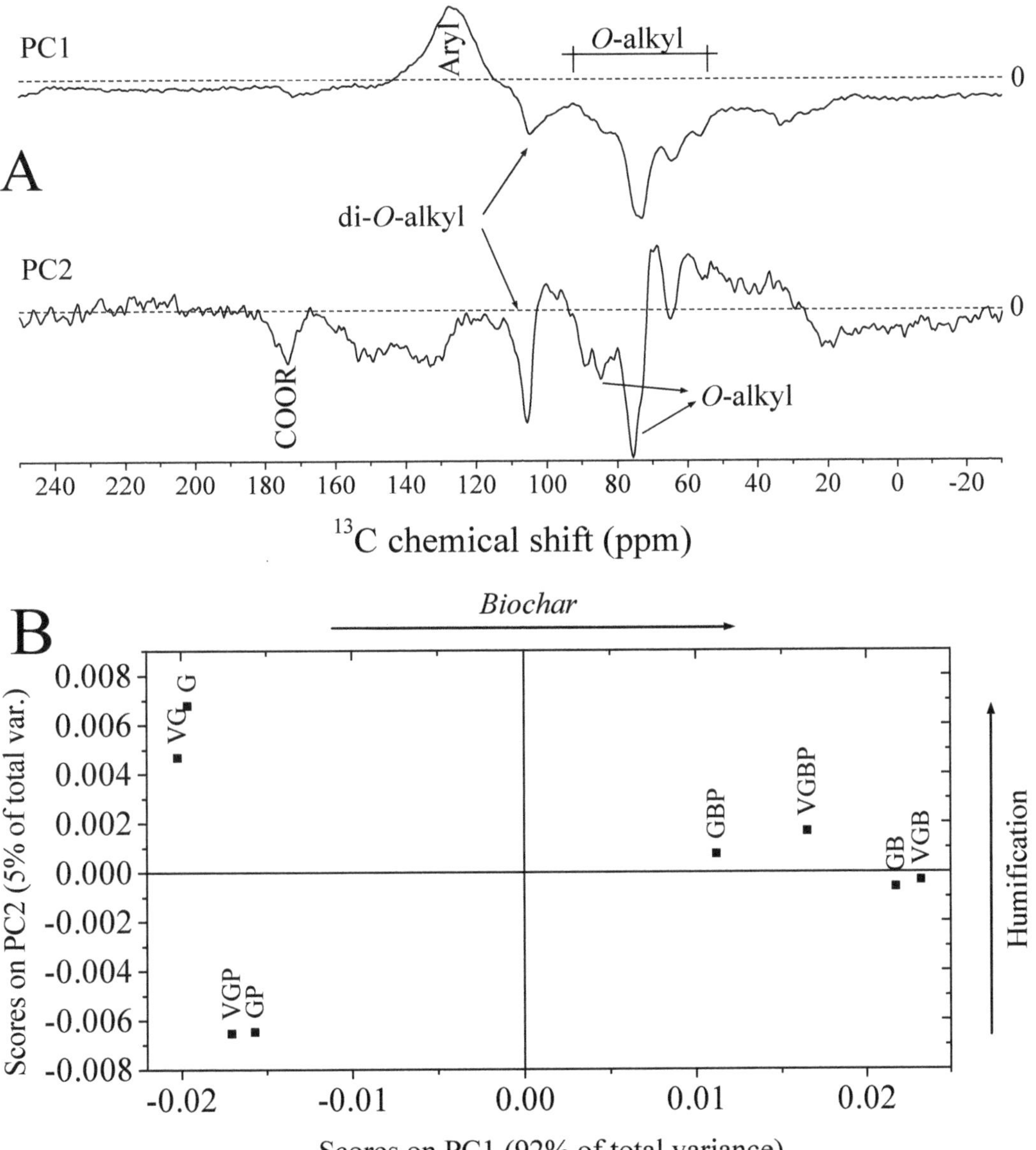

Figure 4. (**A**) PCA loadings obtained from the ^{13}C NMR spectra. (**B**) PCA scores. G: grass compost; VG: grass vermicompost; GP: grass + phosphate compost; VGP: grass + phosphate vermicompost; GBP: grass + biochar + phosphate compost; VGBP: grass + biochar + phosphate vermicompost.

The ^{31}P NMR and its multivariate curve resolution analysis (Figure 5) indicate that, in the absence of biochar, the addition of phosphate to the compost significantly changed the distribution of P species, with a likely predominance of pyrophosphate (with a chemical shift of approximately −1.3 ppm). Moreover, in samples without phosphate or with phosphate and biochar, the predominant form was likely organic (mono and di-ester phosphate, with chemical shifts of approximately 0.96 ppm and 2 ppm, respectively) or mixed with non-hydrolysed monobasic phosphate (chemical shift of 0.96 ppm).

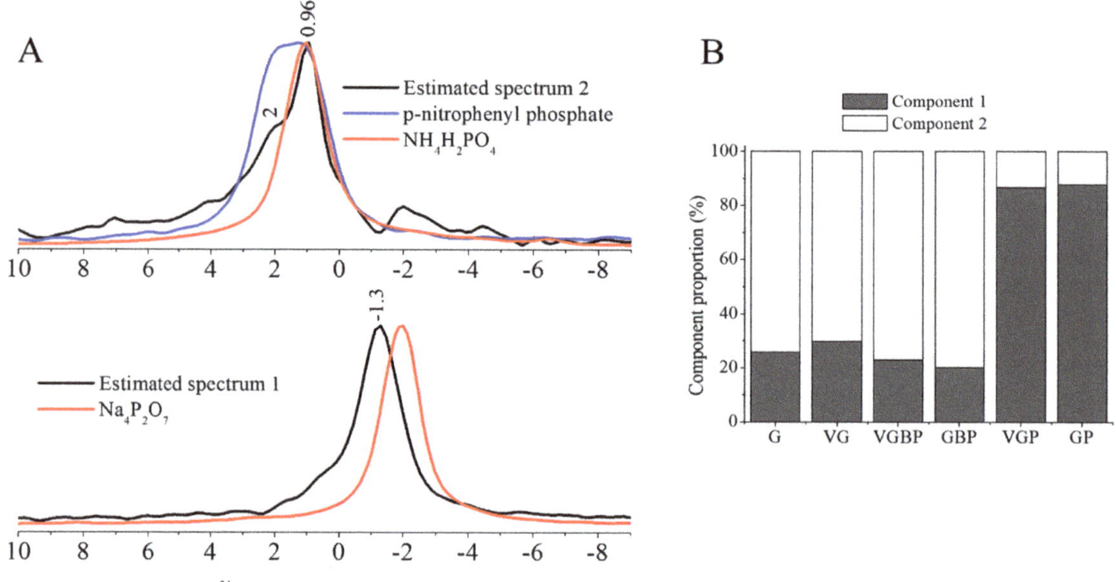

Figure 5. Multivariate curve resolution results obtained from the ^{31}P NMR spectra showing the two component solutions. (**A**): estimated component spectra; discontinued lines show the ^{31}P NMR spectra of ^{31}P standards. (**B**): estimated concentrations. G: grass compost; VG: grass vermicompost; GP: grass + phosphate compost; VGP: grass + phosphate vermicompost; GBP: grass + biochar + phosphate compost; and VGBP: grass + biochar + phosphate vermicompost.

4. Discussion

The grass clipping compost obtained from the International Airport of Rio de Janeiro did not exhibit limitations for composting, which was consistent with many other works [27,28]. We found that grass clippings could also be used for vermicomposting because a statistically significant increase of 400% of the earthworm population was observed during the studied period (Figure 1a). Curry and Schmidt [30] cited, amongst a wide range of organic materials, the presence of grass in the digestive tract of these soil animals, which indicates that grass clipping material can be used to feed earthworms and then in vermicomposting.

However, when SSP was added to the composting substrate, a detrimental effect was observed. The acid reaction from the sulphate present in the phosphate inhibited microbial activity, with the resulting compost exhibiting lower N concentrations, higher C/N ratios (Figure 2b) and higher levels of labile structures (uronic acids—Figure 4). Moreover, the phosphate severely inhibited earthworm reproduction and led to a significant decrease in the earthworm population (i.e., death or escape of the earthworms, Figure 1a). The phosphate dosage (1 kg m^{-3}) used in the present experiment resulted in a significant increase in compost acidity (pH < 5) and in the ionic strength of the substrates (Figure 1b,c), which are factors that negatively affect the development and reproduction of earthworms of this genus [31]. This acidification may result from the conversion of applied phosphate into pyrophosphate in the absence of biochar, as detected by ^{31}P NMR (Figure 5). On the other hand, in the absence of phosphate or in the presence of biochar, the predominant forms of P were organic or non-hydrolysed monobasic phosphate, indicating the highest biological activity and incorporation of P into biological tissues.

In addition, the presence of certain metals (Cd, Cu, Ni, As, Cr and Pb) and radionuclides (^{226}Ra, ^{228}Ra and ^{210}Pb) has been reported in phosphate fertilisers commercialised in Brazil [32], albeit in reduced concentrations. However, their solubility may have increased because of the pronounced pH decrease, thus contributing to the detrimental effect on the earthworms. Nevertheless, additional specific studies are required to confirm this hypothesis. It should be noted that in a review on the uptake and accumulation of heavy metals by earthworms, the metal concentration in the soil was found to be a poor predictor of metal concentration in the earthworms' tissues, and pH was considered to be the main factor associated with metal uptake by earthworms [33].

The presence of biochar mitigated these detrimental effects of phosphate. The pyrolysed biomass acquired different properties depending on the pyrolysis temperature and nature of the feedstock [13,34]. Biochar application has been found to promote changes in nutrient availability (usually pH increase, N immobilisation etc.) [19]; interference with organism and plant signalling compounds; xenobiotic detoxification; and refuge availability for pathogens or growth-promoting organisms, etc. These changes are examples of multiple beneficial or detrimental effects (direct and indirect) of biochar on soil quality and plant production [24,35–37].

The litter earthworm is sensitive to high acidity, high salinity and certain toxic elements, and the treatments with biochar exhibited increased pH and decreased EC, similar or close to the values from the treatments without phosphate. These characteristics may have contributed to the mitigation of detrimental effects of phosphate on the earthworm population and also towards microbiological activity, as indicated by the humification of the biomass (Figure 4).

The ^{31}P NMR spectra and their MCR analysis (Figure 5) show that, in the absence of biochar, the addition of phosphate to the compost significantly changed the distribution of P species, with a likely predominance of pyrophosphate. Moreover, in samples without phosphate or with phosphate and biochar, the predominant form was likely organic (mono and di-ester phosphate) or mixed with monobasic phosphate. This conversion of applied phosphate into pyrophosphate in the absence of biochar may explain the strong acidification of the medium as well as the increase in EC. Biochar was able to prevent this conversion and favoured the formation of P organic species, which may have been caused by the highest microbial activity in the presence of biochar as indicated by the highest compost humification (Figure 3).

Vermicomposting increased the N concentration of the tested substrates (Figure 2c), probably due to the labile organic matter evolving to CO_2 and resulting in a relative enrichment in N.

Biochar addition to the substrates led to lower N concentrations in the compost and vermicompost (Figure 2b), which may have been caused by a dilution effect related to the inclusion of material with high recalcitrant C concentrations and low N concentrations.

A decrease in the C/N atomic ratios was observed after vermicomposting (Figure 2c), indicating a further humification process promoted by the earthworms. The relative increase of N concentrations likely occurred because the raw materials were N-poor and cellulose degradation (CO_2 emission) was favoured by the presence of earthworms; we confirmed this using the ^{13}C NMR spectra (Figures 3 and 4).

5. Conclusions

A detrimental effect of SSP on macro- and micro-biota was observed with a drastic decrease in earthworm population and lower compost and vermicompost humification.

This detrimental effect of SSP was likely caused by acidic conditions resulting from pyrophosphate formation and high salinity, inhibiting humification. This was indicated by the presence of materials with increased labile structures, low N concentrations and high C/N ratios. However, biochar mitigated this negative effect during composting by maintaining microbial activity, which was indicated by higher humification, and during vermicomposting, by mitigating the detrimental effect of SSP on the earthworm population. Biochar probably reduces the deleterious acidity and salinity induced by single superphosphate by decreasing pyrophosphate formation, avoiding acidification and salinity.

SSP alone is not recommended for the composting and vermicomposting of grass clippings, but in combination with biochar it should be tested as a soil conditioner, on account of its greater proportion of stabilized C (from grass clippings and pyrogen) and organic P.

Author Contributions: Conceptualisation (E.H.N.; F.d.C.B.; V.d.M.B. and H.L.d.C.C.); methodology (E.H.N.; F.d.C.B.; R.A. and V.d.M.B.); investigation (E.H.N.; F.d.C.B.; R.A. and V.d.M.B.); formal analysis (E.H.N.); visualisation (E.H.N.); writing—original draft, review and editing (E.H.N.; R.A.; F.d.C.B. and H.L.d.C.C.); supervision and funding acquisition (E.H.N.; F.d.C.B.; V.d.M.B. and H.L.d.C.C.); project administration (F.d.C.B.). All authors have read and agreed to the published version of the manuscript.

Funding: Funding was provided by: the Funding Authority for Studies and Projects (Financiadora de Estudos e Projetos—FINEP) grant project "Production of organic fertilisers and substrates from residues of grass maintenance in urban areas"; the National Counsel of Technological and Scientific Development (Conselho Nacional de Desenvolvimento Científico e Tecnológico—CNPq) for Novotny's and Balieiro's research fellowships (309391/2020-2 and 307434/2020-6, respectively); the Brazilian Federal Agency for Support and Evaluation of Graduate Education (Coordenação de Aperfeiçoamento de Pessoal de Nível Superior—CAPES) for Auccaise's post-doctoral research fellowships (PNPD Nº 03020/09-6); and Fundação de Amparo à Pesquisa do Estado do Rio de Janeiro (FAPERJ) for Novotny's research fellowship (E-26/202.874/2018).

Institutional Review Board Statement: Not applicable since the work does not involve humans or animals.

Informed Consent Statement: Not applicable.

Data Availability Statement: The data that support the findings of this study are available from the corresponding author upon reasonable request.

Acknowledgments: The authors wish to thank the Nuclear Magnetic Resonance Laboratory of the Brazilian Centre of Physics Research (Laboratório de Ressonância Magnética Nuclear do Centro Brasileiro de Pesquisas Físicas). The authors wish to thank the Embrapa Soils trainees (scholarship granted - PIBIC/Embrapa Solos): João Paulo Moura Barata, Carolina Araújo de Queiroz Costa and Natália Evaristo Belo do Santos, as well as the airport workers who helped conduct the experiments at the airport and at Embrapa Soils. I thank Adriana Aquino from Embrapa Agrobiology for the donation of earthworms and recommendations for the management of the experiment.

Conflicts of Interest: The authors declare no conflict of interest.

References

1. Glaser, B.; Balashov, E.; Haumaier, L.; Guggenberger, G.; Zech, W. Black carbon in density fractions of anthropogenic soils of the Brazilian Amazon region. *Org. Geochem.* **2000**, *31*, 669–678. [CrossRef]
2. Cunha, T.J.F.; Madari, B.E.; Benites, V.M.; Canellas, L.P.; Novotny, E.H.; Moutta, R.O. Chemical fractionation of organic matter and humic acid characterization in anthropogenic dark earth soils of the Brazilian Amazon region. *Acta Amazon.* **2007**, *37*, 91–98. [CrossRef]
3. Novotny, E.H.; Hayes, M.H.B.; Madari, B.; Bonagamba, T.J.; Azevedo, E.R.; Souza, A.A.; Song, G.; Nogueira, C.M.; Mangrich, A.S. Lessons from the Terra Preta de Índios of the Amazon region for the utilisation of charcoal for soil amendment. *J. Braz. Chem. Soc.* **2009**, *20*, 1003–1010. [CrossRef]
4. Novotny, E.H.; Maia, C.M.B.F.; Carvalho, M.T.M.; Madari, B.E. Biochar: Pyrogenic carbon for agricultural use—A critical review. *Rev. Bras. Cienc. Solo* **2015**, *39*, 321–344. [CrossRef]

5.   Glaser, B.; Haumaier, L.; Guggenberger, G.; Zech, W. The 'Terra Preta' phenomenon: A model for sustainable agriculture in the humid tropics. *Naturwissenschaften* **2001**, *88*, 37–41. [CrossRef]

6.   Novotny, E.H.; Azevedo, E.R.; Bonagamba, T.J.; Cunha, T.J.F.; Madari, B.E.; Benites, V.M.; Hayes, M.H.B. Studies of the compositions of humic acids from Amazonian dark earth soils. *Environ. Sci. Technol.* **2007**, *41*, 400–405. [CrossRef]

7.   Steiner, C.; Glaser, B.; Teixeira, W.G.; Lehmann, J.; Blum, W.E.H.; Zech, W. Nitrogen retention and plant uptake on a highly weathered central Amazonian Ferralsol amended with compost and charcoal. *J. Plant Nutr. Soil Sci.* **2008**, *171*, 893–899. [CrossRef]

8.   Noguera, D.; Rondón, M.; Laossi, K.; Hoyos, V.; Lavelle, P.; Cruz de Carvalho, M.H. Contrasted effect of biochar and earthworms on rice growth and resource allocation in different soils. *Soil Biol. Biochem.* **2010**, *42*, 1017–1027. [CrossRef]

9.   Ogawa, M.; Okimori, Y. Pioneering works in biochar research, Japan. *Aust. J. Soil Res.* **2010**, *48*, 489–500. [CrossRef]

10. Kookana, R.S.; Sarmah, A.K.; van Zwieten, L.; Krull, E.; Singh, B. Biochar application to soil: Agronomic and environ-mental benefits and unintended consequences. *Adv. Agron.* **2011**, *112*, 103–143. [CrossRef]

11. Petter, F.A.; Madari, B.E.; Silva, M.A.S.D.; Carneiro, M.A.C.; Carvalho, M.T.D.M.; Marimon Júnior, B.H.; Pacheco, L.P. Soil fertility and upland rice yield after biochar application in the Cerrado. *Pesqui. Agropecu. Bras.* **2012**, *47*, 699–706. [CrossRef]

12. Saito, M.; Marumoto, T. Inoculation with arbuscular mycorrhizal fungi: The status quo in Japan and future prospects. *Plant Soi* **2002**, *244*, 273–279. [CrossRef]

13. Sanpa, S.; Imaki, K.; Sumiyoshi, S.; Shibata, A.; Matsumiya, Y.; Kubo, M. Effect of charcoal from woody waste on the soil bacterial biomass and its plant-growth promoting effect. *Wood Carbonization Res.* **2006**, *2*, 37–42.

14. Lehmann, J.; Rillig, M.C.; Thies, J.; Masiello, C.A.; Hockaday, W.C.; Crowley, D. Biochar effects on soil biota—A review. *Soil Biol. Biochem.* **2011**, *43*, 1812–1836. [CrossRef]

15. Lavelle, P.; Barros, E.; Blanchart, E.; Brown, G.; Desjardins, T.; Mariani, L. SOM management in the tropics. Why feed the soil macrofauna? *Nutr. Cycl. Agroecos.* **2001**, *61*, 53–61. [CrossRef]

16. Whiston, R.A.; Seal, K.J. The occurrence of cellulases in the earthworm *Eisenia foetida*. *Biol. Wastes* **1988**, *25*, 239–242. [CrossRef]

17. Dores-Silva, P.R.; Landgraf, M.D.; Rezende, M.O.O. Chemical monitoring of vermicomposting from domestic sewage sludge. *Quim. Nova* **2011**, *34*, 956–961. [CrossRef]

18. Aquino, A.M.; Oliveira, A.M.G.; Loureiro, D.C. Integrating composting and vermicomposting on the recycling of domestic organic waste. *Embrapa Agrobiol.-Circ. Técnica* **2005**, *12*, 1–4.

19. Giannopoulos, G.; Pulleman, M.M.; Van Groenigen, J.W. Interactions between residue placement and earthworm ecological strategy affect aggregate turnover and N_2O dynamics in agricultural soil. *Soil Biol. Biochem.* **2010**, *42*, 618–625. [CrossRef]

20. Abdul Rida, A.M.M.; Bouché, M.B. The eradication of an earthworm genus by heavy metals in Southern France. *Appl. Soil Ecol.* **1995**, *2*, 45–52. [CrossRef]

21. Li, M.; Liu, Z.; Xu, Y.; Cui, Y.; Li, D.; Kong, Z. Comparative effects of Cd and Pb on biochemical response and DNA damage in the earthworm *Eisenia fetida* (*Annelida, Oligochaeta*). *Chemosphere* **2009**, *74*, 621–625. [CrossRef]

22. Hartenstein, R. Effect of aromatic compounds, humic acids and lignins on growth of the earthworm *Eisenia foetida*. *Soil Biol. Biochem.* **1982**, *14*, 595–599. [CrossRef]

23. Loureiro, S.; Soares, A.M.; Nogueira, A.J. Terrestrial avoidance behaviour tests as screening tool to assess soil contamination. *Environ. Pollut.* **2005**, *138*, 121–131. [CrossRef] [PubMed]

24. International Organization for Standardization. Soil Quality—Avoidance Test for Determining the Quality of Soils and Effects of Chemicals on Behaviour—Part 1: Test with Earthworms (Eisenia fetida and Eisenia andrei), Geneva. 2008. Available online: https://www.sis.se/api/document/preview/910491/ (accessed on 11 December 2019).

25. Sarathchandra, S.U.; Lee, A.; Perrott, K.W.; Rajan, S.S.S.; Oliver, E.H.A.; Gravett, I.M. Effects of phosphate fertilizer applications on microorganisms in pastoral soil. *Aust. J. Soil Res.* **1993**, *31*, 299–309. [CrossRef]

26. Abbiramy, K.S.K.; Ross, P.R.; Paramanandham, J.P. Assessment of acute toxicity of superphosphate to *Eisenia fetida* using paper contact method. *Asian J. Plant Sci. Res.* **2013**, *3*, 112–115.

27. Nakasaki, K.; Hiraoka, S.; Nagata, H. A new operation for producing disease-suppressive compost from grass clippings. *Appl. Environ. Microbiol.* **1998**, *64*, 1015–4020. [CrossRef] [PubMed]

28. Benites, V.M.; Bezerra, F.B.; Mouta, R.O.; Assis, I.R.; Santos, R.C.; Conceição, M. Production of organic fertilisers from composting of waste from the maintenance of grass areas of the Rio de Janeiro International airport. In *Boletim de Pesquisa e Desenvolvimento*; Embrapa-CNPS: Rio de Janero, Brazil, 2004.

29. SAS Institute Inc. *SAS®Visual Statistics 7.1: User's Guide*; SAS Institute Inc.: Cary, NC, USA, 2014.

30. Curry, J.P.; Schmidt, O. The feeding ecology of earthworms—A review. *Pedobiologia* **2006**, *50*, 463–477. [CrossRef]

31. Kaplan, D.L.; Hartenstein, R.; Neuhauser, E.F.; Malecki, M.R. Physicochemical requirements in the environment of the earthworm *Eisenia foetida*. *Soil Biol. Biochem.* **1980**, *12*, 347–352. [CrossRef]

32. Saueia, C.H.R.; Le Bourlegat, F.M.; Mazzilli, B.P.; Fávaro, D.I.T. Availability of metals and radionuclides present in phosphogypsum and phosphate fertilizers used in Brazil. *J. Radioanal. Nucl. Chem.* **2013**, *297*, 189–195. [CrossRef]

33. Nahmani, J.; Hodson, M.E.; Black, S. A review of studies performed to assess metal uptake by earthworms. *Environ. Pollut.* **2007**, *145*, 402–424. [CrossRef]

34. Czimczik, C.I.; Masiello, C.A. Controls on black carbon storage in soils. *Glob. Biogeochem. Cycles* **2007**, *21*, GB3005. [CrossRef]

35. Rondon, M.A.; Lehmann, J.; Ramírez, J.; Hurtado, M. Biological nitrogen fixation by common beans (*Phaseolus vulgaris* L.) increases with bio-char additions. *Biol. Fert. Soils* **2007**, *43*, 699–708. [CrossRef]

36. Warnock, D.D.; Lehmann, J.; Kuyper, T.W.; Rillig, M.C. Mycorrhizal responses to biochar in soil—Concepts and mechanisms. *Plant Soil* **2007**, *300*, 9–20. [CrossRef]
37. Anderson, C.R.; Condron, L.M.; Clough, T.J.; Fiers, M.; Stewart, A.; Hill, R.A. Biochar induced soil microbial community change: Implications for biogeochemical cycling of carbon, nitrogen and phosphorus. *Pedobiologia* **2011**, *54*, 309–320. [CrossRef]

 *agriculture*

Article

Trichoderma Bio-Fertilizer Decreased C Mineralization in Aggregates on the Southern North China Plain

Lixia Zhu, Mengmeng Cao, Chengchen Sang, Tingxuan Li, Yanjun Zhang, Yunxia Chang and Lili Li *

College of Life Science and Agronomy, Zhoukou Normal University, Zhoukou 466001, China;
20181006@zknu.edu.cn (L.Z.); 202007030031@zknu.cn (M.C.); 202007030055@zknu.cn (C.S.);
202007030014@zknu.edu.cn (T.L.); 202007030047@zknu.cn (Y.Z.); 20061015@zknu.edu.cn (Y.C.)
* Correspondence: 19961011@zknu.edu.cn

Abstract: *Trichoderma* bio-fertilizer is widely used to improve soil fertility and carbon (C) sequestration, but the mechanism for increasing C accumulation remains unclear. In this study, effects of *Trichoderma* bio-fertilizer on the mineralization of aggregate-associated organic C were investigated in a field experiment with five treatments (bio-fertilizer substitute 0 (CF), 10% (BF10), 20% (BF20), 30% (BF30) and 50% (BF50) chemical fertilizer nitrogen (N)). Aggregate fractions collected by the dry sieving method were used to determine mineralization dynamics of aggregate-associated organic C. The microbial community across aggregate fractions was detected by the phospholipid fatty acid (PLFA) method. The results indicated that *Trichoderma* bio-fertilizer increased organic C stock across aggregate fractions and bulk soil compared with CF. Cumulative mineralization of aggregate-associated organic C increased with the increasing bio-fertilizer application rate. However, the proportion of organic mineralized C was lower in the BF20 treatment except for <0.053 mm aggregate. Moreover, the PLFAs and fungal PLFA/bacterial PLFA first increased and then decreased with increasing bio-fertilizer application rates. Compared with CF, the increases of bacteria PLFA in >2 mm aggregate were 79.7%, 130.0%, 141.0% and 148.5% in BF10, BF20, BF30 and BF50, respectively. Similarly, the PLFAs in 0.25–2, 0.053–0.25 and <0.053 mm aggregates showed a similar trend to that in >2 mm aggregate. Bio-fertilizer increased the value of fungi PLFA/bacteria PLFA but decreased G+ PLFA/G− PLFA, and BF20 shared the greatest changes. Therefore, appropriate *Trichoderma* bio-fertilizer application was beneficial to improving soil micro-environment and minimizing risks of soil degradation.

Keywords: *Trichoderma* bio-fertilizer; soil aggregate stability; mineralization; microbial community

Citation: Zhu, L.; Cao, M.; Sang, C.; Li, T.; Zhang, Y.; Chang, Y.; Li, L. *Trichoderma* Bio-Fertilizer Decreased C Mineralization in Aggregates on the Southern North China Plain. *Agriculture* **2022**, *12*, 1001. https://doi.org/10.3390/agriculture12071001

Academic Editors: Yinglong Chen, Masanori Saito and Etelvino Henrique Novotny

Received: 9 June 2022
Accepted: 8 July 2022
Published: 11 July 2022

Publisher's Note: MDPI stays neutral with regard to jurisdictional claims in published maps and institutional affiliations.

1. Introduction

Increasing carbon (C) sequestration in cropland has been recognized as an effective way to reduce CO_2 emissions, improve soil structure and promote soil microbial diversity [1,2]. Though protected from microbial decomposition [3], organic C in each aggregate fraction could be affected by agronomic practices (i.e., bio-fertilizer application). Furthermore, the inherent heterogeneity of microbes existing in different aggregate fractions was also influenced by organic material application [2]. Generally, soil aggregate fractions would affect the soil microbial diversity, while soil C contributed to the microbial community structure [4,5]. However, sequestration of soil organic carbon (SOC) was closely related to the C input and output, which were affected by the microbial mineralization of organic materials. Clay soil with high SOC content was more vulnerable to decomposition, showing that dynamics of SOC might also be affected by soil texture which was influenced by aggregate fraction [6]. Therefore, evaluating the sequestration and mineralization of organic C in aggregate fractions and its related microbial community is vital for a deeper understanding of soil C stability.

Studies have demonstrated that long-term fertilization significantly affected SOC stability in different aggregate fractions [7–9], and mineralization of organic C would vary with soil aggregates [6,10]. The mineralization of C in small aggregates was vulnerable to soil conditions in an experiment in which soil types concluded Alfisols, Inceptisols, Mollisols and Ultisols [11]. Reeves et al. [12] showed that the mineralization of C in micro-aggregates was higher in Vertisol, while Wang et al. [13] found an opposite trend in the soil of Udic Ferralsols. No difference in organic C mineralization between different aggregate fractions was also observed in Dermosols [14]. Such diverse results clearly indicated that mineralization of aggregate-associated organic C varied with soil profiles. Soil microbes, the driving force of organic C decomposition, were affected by soil textures and aggregate class [13,15]. Therefore, it is important to investigate changes in soil microbial community across aggregate fractions to deeply understand soil C stability. Generally, organic materials clearly increased fungi decomposing biosolids and decreased the abundance of CO_2-emission-related fungi [16]. However, little is known about variations of soil microbial community in aggregate fractions and how different classes of aggregates protect and sequester C in soil cultivated with *Trichoderma* bio-fertilizer.

The winter wheat-summer maize cropping system is the major planting pattern in the North China Plain (NCP), which occupies around 16 M ha [17]. The Shajiang Calci-Aquic Vertosol, an important soil resource, is widely distributed in the southern NCP. However, research information is lacking about the mineralization of aggregate-associated organic C in this area. Thus, this study was conducted to investigate (1) the aggregate distribution and aggregate-associated organic C under the dry sieving method, (2) the mineralization dynamics of C in aggregates in soil treated with bio-fertilizer and (3) the soil microbial community within aggregate fractions in Shajiang Calci-Aquic Vertosol. We hope this research will provide fundamental evidence to strengthen the understanding of microbial regulation of C dynamics across aggregate fractions and guide appropriate bio-fertilizer application in the southern NCP.

2. Materials and Methods

2.1. Study Area and Experimental Design

The study was conducted in Xiping county in 2018 in the southern NCP, China (N 113°12′, E 33°27′, with an altitude of 64 m above sea level), with a mean annual precipitation of 786 mm. The field had been cultivated for more than 50 years before this study started. The mean annual temperature is 14.7 °C, and the sunshine duration has been 2181 h over the past 50 years. Based on the Chinese Soil Taxonomy, the soil in this area is classified as a Shajiang Calci-Aquic Vertosol with a pH value of 6.9. The soil bulk density, SOC, total nitrogen (TN), mineral nitrogen, available phosphorus and available potassium of the 0–20 cm depth are 1.31 g cm^{-3}, 9.4 g kg^{-1}, 0.88 g kg^{-1}, 14.38 mg kg^{-1}, 6.06 mg kg^{-1} and 179.23 mg kg^{-1}, respectively.

Five treatments were included in this experiment: CF (100% chemical fertilizer N), BF10 (chemical fertilizer N supplemented with 10% *Trichoderma* bio-fertilizer N), BF20 (chemical fertilizer N supplemented with 20% *Trichoderma* bio-fertilizer N), BF30 (chemical fertilizer N supplemented with 30% *Trichoderma* bio-fertilizer N) and BF50 (chemical fertilizer N supplemented with 50% *Trichoderma* bio-fertilizer N). Each treatment received the same amounts of N, P_2O_5 and K_2O (220, 90 and 90 kg ha^{-1}, respectively) during the maize growing season. The chemical fertilizers were urea (N 46%), calcium super phosphate (P_2O_5 12%) and potassium sulphate (K_2O 51%). The nitrogen, phosphorus and potassium contents of *Trichoderma* bio-fertilizer were deducted and supplemented with corresponding chemical fertilizer in each treatment. The details of the *Trichoderma* bio-fertilizer were listed in Zhu et al. [9]. Fermented with wheat straw, the bio-fertilizer had approximately 2×10^8 colony forming units per gram of *Trichoderma asperellum* ACCC30536, with 297.2 g kg^{-1} organic matter and 120.1 g kg^{-1} total nitrogen. Both the *Trichoderma* bio-fertilizer and chemical fertilizer were spread evenly on the surface of each plot and thoroughly mixed with the top 0–20 cm layer by a rotary cultivator a week before sowing the maize. Summer maize in each plot was planted at a density of 65,000 plants ha^{-1} in early June and harvested in early October each year.

2.2. Soil Sampling and Analysis

Soil cores were sampled after the maize harvest in October 2020. Five random soil cores from the top 20 cm layer of each plot were taken and mixed thoroughly as one composite sample. Visible materials were removed and stored at 4 °C before analysis. Subsamples of 300 g were shaken on a motorized vibratory sieve-shaker (8411; Zhejiang Chenxin Machine Equipment Co., Ltd., Shaoxing, China) for 3 min with a mesh size of 2, 0.25 and 0.053 mm to obtain four size fractions: >2, 0.25–2, 0.053–0.25 and <0.053 mm.

The incubation experiment was prepared as follows: 30 g samples of the four aggregate fractions were incubated in 500 mL buckets at 25 °C and at 60% water holding capacity for 60 days. The released CO_2 was trapped in NaOH solution and determined after 5, 10, 15, 20, 30 and 60 days. The trapped CO_2 was precipitated with 1.0 mol L^{-1} $BaCl_2$ and then titrated with standard 0.1 mol L^{-1} HCl to quantify the released CO_2. Cumulative mineralized CO_2-C was used for comparison between aggregate size classes and fertilization treatments.

The PLFAs were extracted and identified according to Bligh and Dyer [18]. Fatty acid methylesters were analyzed using an Agilent 6850 system (Agilent Technologies, Palo Alto, CA, USA), equipped with a DP-5MS capillary column (25 m × 200 μm, i.d. 0.33 μm). Assignment of microbial categories to the various PLFA biomarkers was based on Luo et al. [19].

2.3. Data Analyses

The aggregate stability was determined by mean weight diameter (MWD), and MWD was calculated with the following equation:

$$MWD = \sum(X_i W_i) \tag{1}$$

Of which X_i was the mean diameter of the class (mm), and W_i was the proportion of each aggregate class in relation to the weight of the soil samples.

Total respiration was calculated with the CO_2-C produced from each aggregate class (CO_2Agg) and the mass of each aggregate class (MAgg) in the total soil mass (SM), according to Xie et al. [15]:

$$\text{Total respiration (mg C kg}^{-1}) = CO_2\text{ Agg} \times \text{MAgg/SM} \times 100 \tag{2}$$

The contribution of organic C mineralization of each aggregate fraction to bulk soil was calculated with the ratio of cumulative CO_2-C production of each aggregate multiplied by the mass of each aggregate fraction/total mineralization of organic C in bulk soil.

2.4. Statistical Analysis

For each of these variables, a mean value was obtained from the results for the three composite samples, significant differences between the means were identified by performing a one-way analysis of variance (ANOVA), and the least significant difference (LSD) was computed to compare the means of the above variables ($p < 0.05$) using the statistical package SPSS 19.0. Statistical correlations between mineralization of organic C and microbial community were assessed with correlation analysis, and Pearson correlation coefficients were presented. Prior to analysis, data were tested to ensure that they met homogeneity of variance and normality requirements. All the figures were accomplished with SigmaPlot 12.0 (Systat Software Inc., London, UK).

3. Results

3.1. Distribution of Soil Aggregates

Soil aggregate distribution was significantly affected by the bio-fertilizer application. As the dominant fraction, >2 mm aggregate accounted for 34.90% in CF treatment and for 48.94% in BF20 treatment. The proportion of 0.25–2 mm aggregate ranged from 26.08% (CF) to 36.60% (BF20) (Table 1). For the >2 mm aggregate, the highest value was recorded in BF20, followed by BF30, BF50, BF10 and CF. Compared with CF, the increases of >2 mm aggregate were 18.70%, 40.22%, 35.97% and 32.75% in BF10, BF20, BF30 and BF50, respectively ($p < 0.05$). No significance was observed among BF20, BF30 and BF50 in >2 mm aggregate. Additionally, bio-fertilizer significantly increased the content of 0.25–2 mm aggregate, while BF20, BF30 and BF50 showed no difference between each other. Differently, the distribution of 0.053–0.25 mm aggregate showed a trend of CF > BF10 > BF20 = BF30 = BF50. Bio-fertilizer significantly decreased the content of <0.053 mm aggregate, and no difference was observed among the four bio-fertilizer treatments. Moreover, bio-fertilizer significantly increased MWD and the increase of MWD in BF20 was the highest (35.6%). *Trichoderma* bio-fertilizer promoted the formation of macro-aggregate (>0.25 mm) in soil but decreased proportions of micro-aggregate (0.053–0.25 mm) and clay (<0.053 mm), and then increased soil aggregate stability.

Table 1. Distribution of soil aggregate and soil aggregate stability assessed by dry sieving method.

Treatment	Distribution of Soil Aggregate (%)				MWD (mm)
	>2 mm	0.25–2 mm	0.053–0.25 mm	<0.053 mm	
CF	34.90 ± 1.34 c	26.08 ± 2.40 c	26.54 ± 1.46 a	12.48 ± 3.92 a	1.04 ± 0.04 d
BF10	41.43 ± 0.90 b	31.71 ± 1.00 b	21.93 ± 1.59 b	4.93 ± 1.95 b	1.22 ± 0.01 c
BF20	48.94 ± 0.96 a	36.60 ± 0.96 a	13.28 ± 0.40 c	1.18 ± 0.33 b	1.41 ± 0.00 a
BF30	47.46 ± 1.00 a	34.29 ± 1.28 a	14.89 ± 1.27 c	3.37 ± 1.01 b	1.36 ± 0.03 b
BF50	46.33 ± 1.04 a	35.43 ± 0.63 a	15.23 ± 0.89 c	3.01 ± 1.15 b	1.35 ± 0.02 b

Notes: Values are means ± standard deviation of three replicate plots. Different lower case letters in the same column indicate significant differences between treatments for the same aggregate fraction ($p < 0.05$). CF: 100% chemical fertilizer N; BF10: chemical fertilizer N supplemented with 10% bio-fertilizer N; BF20: chemical fertilizer N supplemented with 20% bio-fertilizer N; BF30: chemical fertilizer N supplemented with 30% bio-fertilizer N; BF50: chemical fertilizer N supplemented with 50% bio-fertilizer N.

3.2. Organic C in Bulk Soil and Aggregate Fractions

As shown in Figure 1, the aggregate-associated organic C was significantly affected by the bio-fertilizer application and aggregate size. Compared with CF, organic C in aggregates of >2, 0.25–2, 0.053–0.25 and <0.053 mm were significantly increased by 11.82%, 11.98%, 10.86% and 11.81% in BF10, respectively. The organic C in aggregates of >2, 0.25–2, 0.053–0.25 and <0.053 mm in BF20 was 18.15%, 18.60%, 17.75%, and 18.46% higher than that in CF, respectively. The increases of organic C in >2, 0.25–2, 0.053–0.25 and <0.053 mm aggregate fractions were 17.85%, 18.08%, 17.15% and 18.04% in BF30, and the increases were 17.99%, 18.31%, 17.15% and 17.74% in BF50, respectively, relative to CF. No difference in aggregate-associated organic C was observed in >2, 0.25–2 and 0.053–0.25 mm aggregate fractions among BF20, BF30 and BF50. Obviously, the organic C in <0.053 mm aggregate in BF50 was significantly lower than that in BF20.

3.3. Mineralization of Organic C in Aggregate

The mineralization dynamics of aggregate-associated organic C were noticeably affected by *Trichoderma* bio-fertilizer application, especially in the first 15 days (Figure 2), and then the mineralization rates were relatively low and remained stable. The BF50 treatment yielded the highest mineralization rate, and the lowest value was recorded in CF treatment across aggregate fractions and bulk soil. The >2 mm aggregate produced the highest amount of CO_2-C, and the aggregate of <0.053 mm yielded the lowest value.

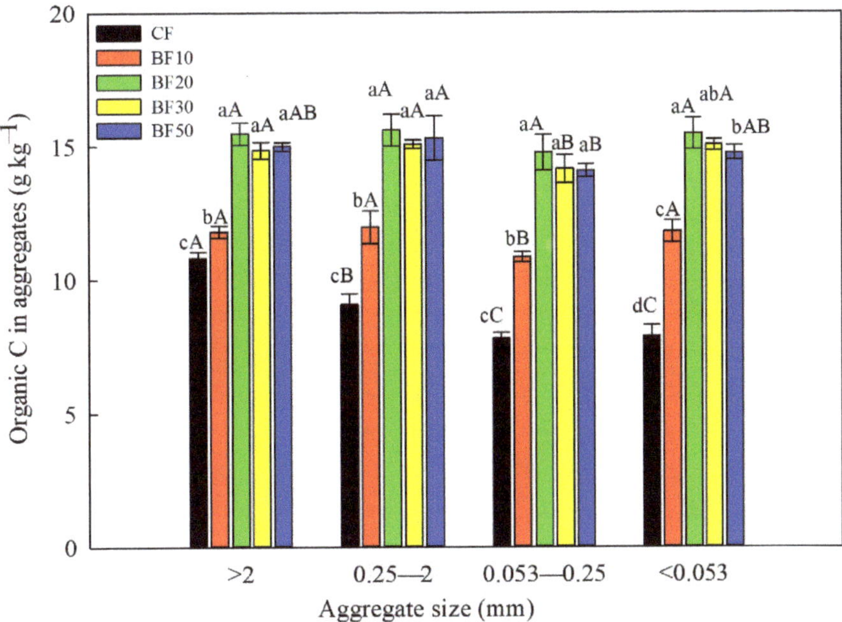

Figure 1. Organic C in different aggregate classes in the five treatments. Different lowercase letters indicate significant differences between treatments, and different uppercase letters indicate differences between aggregate size classes at $p < 0.05$. Error bars represent standard deviation of means ($n = 3$). CF: 100% chemical fertilizer N; BF10: chemical fertilizer N supplemented with 10% bio-fertilizer N; BF20: chemical fertilizer N supplemented with 20% bio-fertilizer N; BF30: chemical fertilizer N supplemented with 30% bio-fertilizer N; BF50: chemical fertilizer N supplemented with 50% bio-fertilizer N.

After 60 days of incubation, the cumulative mineralization of aggregate-associated organic C in the five treatments differed from each other (Figure 3a). The >2 mm aggregate shared the highest CO_2-C among the five treatments, while 0.25–2 and 0.053–0.25 mm aggregate fractions produced similar amounts of CO_2-C. The CO_2-C yielded from the <0.053 mm aggregate was the lowest among the four aggregate fractions. Compared with CF, the highest values of CO_2-C were recorded in the BF50 treatment across aggregate fractions. Increases of CO_2-C in BF50 were 121.00% (>2 mm), 21.97% (0.25–2 mm), 35.18% (0.053–0.25 mm) and 77.66% (<0.053 mm) relative to CF, respectively. The bulk soil respired 130.28, 188.35, 244.37, 288.16 and 314.59 mg C kg^{-1} soil in CF, BF10, BF20, BF30 and BF50, respectively (Figure 3b). Further analysis showed that the mineralization rate of organic C was generally similar for CF and BF10, while a significantly lower value was observed in BF20 relative to CF (Table 2). There was no difference between BF20, BF30 and BF50 in aggregate fractions of 0.25–2, 0.053–0.25 and <0.053 mm. However, the highest mineralization rate was observed in BF50 treatment in bulk soil.

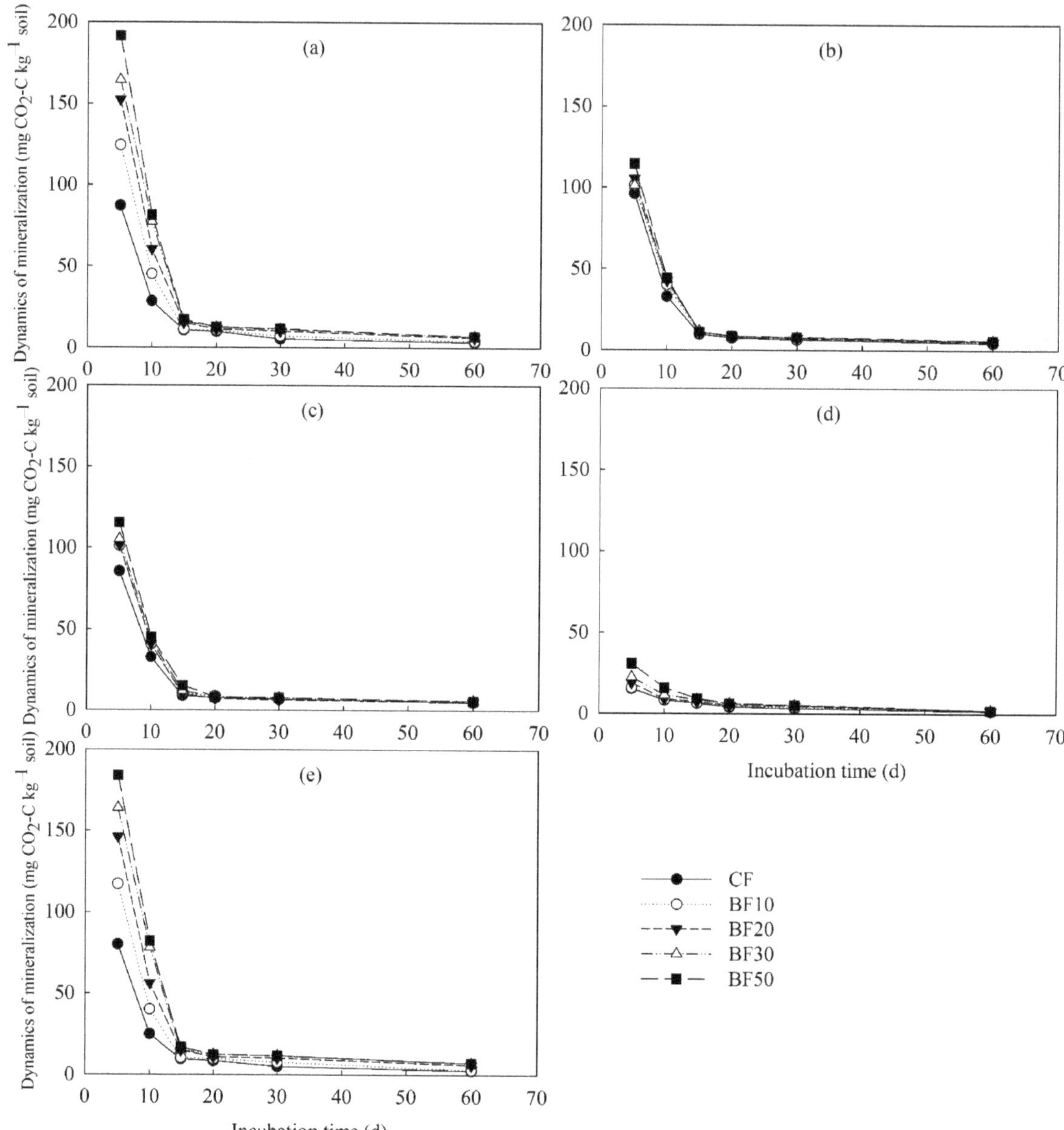

Figure 2. The dynamics of organic C of mineralization in >2 mm (**a**), 0.25–2 mm (**b**), 0.053–0.25 mm (**c**), <0.053 mm aggregate fractions (**d**) and bulk soil (**e**) for the five treatments. Values are means of three replicates. CF, BF10, BF20, BF30 and BF50 are defined as in Figure 1.

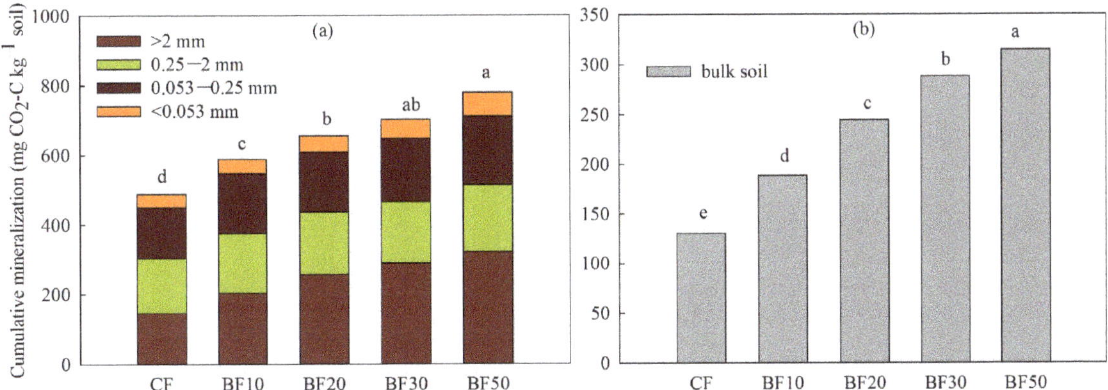

Figure 3. Cumulative mineralization of organic C in different aggregate fractions (**a**) and bulk soil (**b**) under the five treatments. Different lowercase letters indicate significant differences among treatments at $p < 0.05$. CF, BF10, BF20, BF30 and BF50 are defined as in Figure 1.

Table 2. The rate of organic C respiration in the four aggregate fractions and bulk soil.

Treatment	Rate of Organic C Respiration (%)				
	>2 mm	0.25–2 mm	0.053–0.25 mm	<0.053 mm	Bulk Soil
CF	1.78 ± 0.12 aA	1.74 ± 0.14 aA	1.61 ± 0.05 aA	0.49 ± 0.12 aC	1.56 ± 0.12 abB
BF10	1.71 ± 0.14 aA	1.44 ± 0.22 aC	1.44 ± 0.24 aC	0.38 ± 0.05 aD	1.54 ± 0.22 abB
BF20	1.41 ± 0.20 bA	0.97 ± 0.13 bB	0.92 ± 0.14 bB	0.26 ± 0.13 aC	1.35 ± 0.11 bA
BF30	1.61 ± 0.11 abA	0.98 ± 0.21 bB	1.00 ± 0.35 bB	0.32 ± 0.05 aC	1.58 ± 0.12 abA
BF50	1.79 ± 0.11 aA	1.05 ± 0.15 bB	1.08 ± 0.12 bB	0.40 ± 0.08 aC	1.72 ± 0.12 aA

Notes: Different lowercase letters within a column indicate significant differences between treatments, and different uppercase letters in the same row indicate differences between aggregate fractions at $p < 0.05$. CF, BF10, BF20, BF30 and BF50 are defined as in Table 1.

3.4. Microbial Community Composition in Aggregate

Trichoderma bio-fertilizer significantly increased the PLFAs of total microbes, bacteria, fungi, actinomycetes, G+ bacteria and G− bacteria across aggregate fractions and bulk soil (Figure 4). Total PLFA and bacteria PLFA in each aggregate fraction and bulk soil increased with increasing bio-fertilizer application rate, and BF50 yielded the highest PLFAs. Compared with CF, increases of bacteria PLFA in >2 mm fraction were 79.7%, 130.0%, 141.0% and 148.5% in BF10, BF20, BF30 and BF50; increases of fungi PLFA were 108.4%, 163.3%, 172.9% and 195.2%. Actinomycetes PLFAs were increased by 20.9%, 20.4%, 29.6% and 54.4% in BF10, BF20, BF30 and BF50, respectively, relative to CF. Increases of G+ PLFA reached 34.0%, 44.1%, 54.1% and 61.1% in BF10, BF20, BF30 and BF50, respectively, relative to CF. Additionally, G− PLFAs were enhanced by 22.5%, 34.3%, 38.8% and 45.5% in BF10, BF20, BF30 and BF50, respectively, in comparison with CF. The PLFAs in 0.25–2, 0.053–0.25 and <0.053 mm aggregate fractions showed a similar trend to that in >2 mm aggregate fraction.

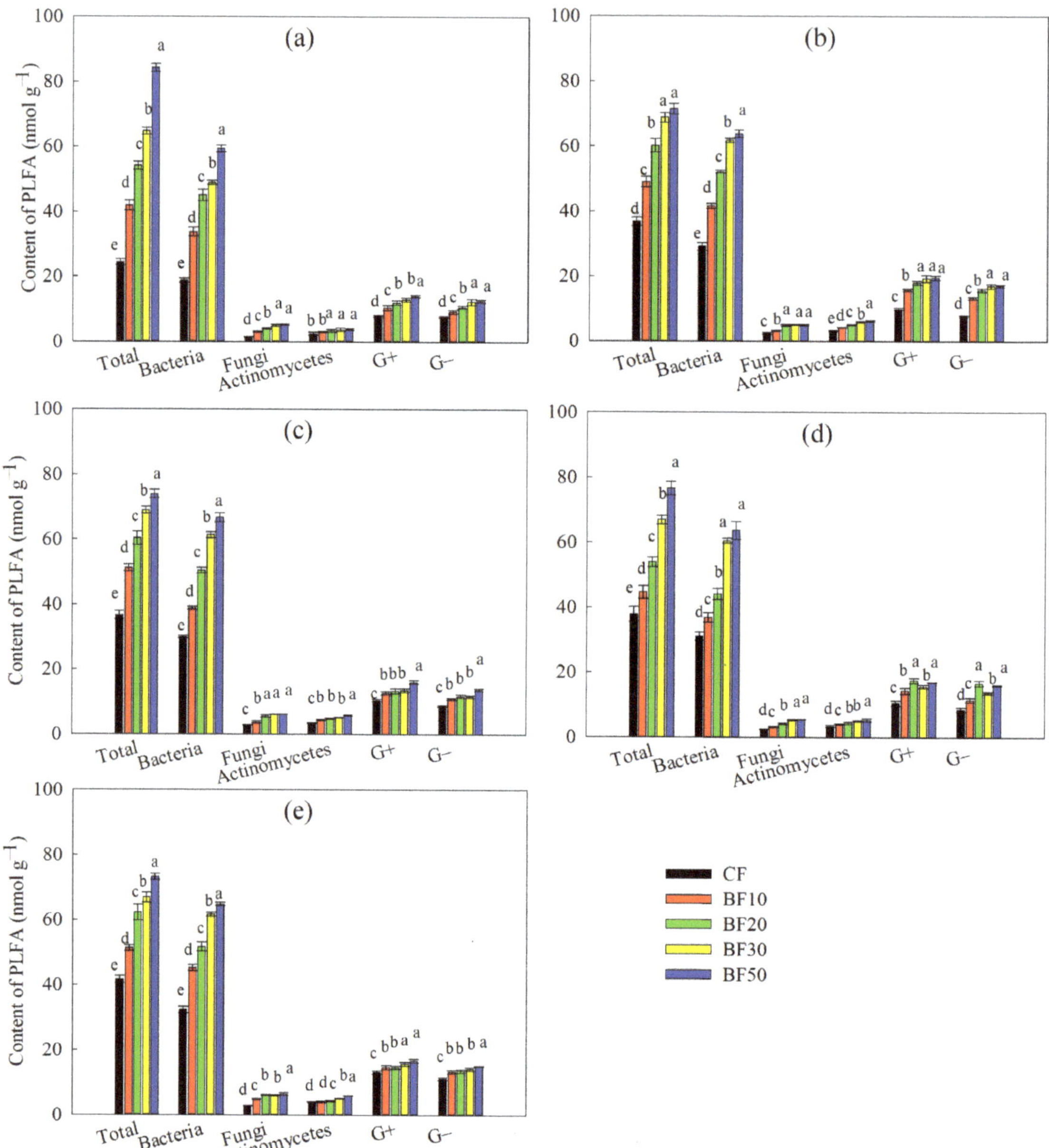

Figure 4. The soil microbial community in >2 mm (**a**), 0.25–2 mm (**b**), 0.053–0.25 mm (**c**), <0.053 mm aggregate fractions (**d**) and bulk soil (**e**) for the five treatments. Error bars represent the standard deviation of the mean ($n = 3$). Different lowercase letters above bars for each PLFA indicate significant differences at $p < 0.05$. CF, BF10, BF20, BF30 and BF50 are defined as in Figure 1.

The soil microbial community diversity was significantly affected by *Trichoderma* bio-fertilizer application and aggregate fractions (Table 3). No significance of Shannon index *H* was observed among treatments in >2 and 0.25–2 mm aggregate fractions. Compared with CF, BF30 and BF50 significantly decreased H by 11.72% and 10.94% in 0.053–0.25 mm aggregate and decreases of H were 8.00% and 4.80% in <0.053 mm in BF30 and BF50 treatments, respectively. Bio-fertilizer significantly decreased values of H in bulk soil, and the lowest H value was observed in BF30. Bio-fertilizer significantly increased the value of fungi PLFA/bacteria PLFA, and the highest value was recorded in BF30 in >2 mm aggregate fraction. Differently, BF20 yielded the highest value of fungi PLFA/bacteria PLFA in 0.25–2, 0.053–0.25 and <0.053 mm aggregates and bulk soil. BF10, BF20 and BF50 significantly increased values of G+ PLFA/G− PLFA, and the increases were 9.71%, 6.80% and 8.74%, respectively, relative to CF. No difference in G+ PLFA/G− PLFA was observed between CF and BF30 in >2 mm aggregate. Notably, decreases of G+ PLFA/G− PLFA in BF20 were the most in 0.25–2, 0.053–0.25 and <0.053 mm aggregate fractions and bulk soil relative to CF.

Table 3. The diversity of soil microbial community in aggregate fractions and bulk soil in different treatments.

Diversity Index	Treatment	Aggregate Fractions (mm)				Bulk Soil
		>2	0.25–2	0.053–0.25	<0.053	
	CF	1.33 ± 0.12 a	1.26 ± 0.01 a	1.28 ± 0.11 a	1.25 ± 0.02 a	1.32 ± 0.01 a
	BF10	1.23 ± 0.11 a	1.25 ± 0.08 a	1.29 ± 0.12 a	1.27 ± 0.03 a	1.24 ± 0.01 b
H	BF20	1.17 ± 0.08 a	1.25 ± 0.02 a	1.22 ± 0.01 a	1.30 ± 0.02 a	1.24 ± 0.01 b
	BF30	1.21 ± 0.08 a	1.20 ± 0.13 a	1.13 ± 0.01 b	1.15 ± 0.02 b	1.16 ± 0.03 c
	BF50	1.14 ± 0.11 a	1.20 ± 0.06 a	1.14 ± 0.01 b	1.19 ± 0.02 b	1.19 ± 0.02 c
	CF	0.078 ± 0.004 c	0.08 ± 0.008 b	0.09 ± 0.006 b	0.08 ± 0.002 c	0.09 ± 0.003 d
	BF10	0.091 ± 0.004 b	0.08 ± 0.005 b	0.10 ± 0.008 b	0.09 ± 0.000 c	0.11 ± 0.002 b
Fungi PLFA/Bacteria PLFA	BF20	0.091 ± 0.002 b	0.09 ± 0.006 a	0.11 ± 0.007 a	0.10 ± 0.002 a	0.12 ± 0.005 a
	BF30	0.102 ± 0.005 a	0.08 ± 0.002 b	0.10 ± 0.003 b	0.09 ± 0.003 b	0.10 ± 0.001 c
	BF50	0.089 ± 0.005 b	0.08 ± 0.003 b	0.09 ± 0.002 b	0.09 ± 0.003 b	0.10 ± 0.006 c
	CF	1.03 ± 0.006 b	1.27 ± 0.015 a	1.20 ± 0.008 a	1.23 ± 0.016 a	1.18 ± 0.023 a
	BF10	1.13 ± 0.016 a	1.18 ± 0.032 b	1.16 ± 0.020 b	1.24 ± 0.008 a	1.10 ± 0.006 b
G+ PLFA/G− PLFA	BF20	1.10 ± 0.021 a	1.14 ± 0.012 c	1.13 ± 0.010 c	1.05 ± 0.020 c	1.08 ± 0.010 b
	BF30	1.06 ± 0.062 b	1.14 ± 0.017 c	1.17 ± 0.016 b	1.15 ± 0.016 b	1.11 ± 0.015 b
	BF50	1.12 ± 0.022 a	1.15 ± 0.011 bc	1.17 ± 0.010 b	1.06 ± 0.003 c	1.11 ± 0.030 b

Notes: Values are means ± standard deviation of three replicate plots. Different lowercase letters within a column for the same index indicated significant differences among treatments at $p < 0.05$. CF, BF10, BF20, BF30 and BF50 are defined as in Table 1.

As shown in Table 4, aggregate-associated organic C was significantly positively correlated with bacteria PLFA ($R = 0.865$, $p < 0.01$) and fungi PLFA ($R = 0.881$, $p < 0.01$), but was significantly negatively correlated with G+ PLFA/G− PLFA ($R = -0.450$, $p < 0.05$) and H ($R = -0.665$, $p < 0.05$). A significant positive correlation was observed between aggregate distribution and fungi PLFA ($R = 0.141$, $p < 0.05$). However, there was no correlation between mineralization of aggregate-associated organic C and microbial community.

Table 4. Relationships between mineralization of organic C and microbial community.

Index	Bacteria PLFA	Fungi PLFA	Fungi PLFA/Bacteria PLFA	G+ PLFA/G− PLFA	H
Organic C in aggregate	0.865 **	0.881 **	0.328	−0.450 *	−0.665 **
Mineralization of C in aggregate	0.215	0.268	0.247	−0.209	−0.395
Aggregate distribution	0.114	0.141 *	0.090	0.180	0.077

Notes: * indicated significance at $p < 0.05$; ** indicated significance at $p < 0.01$.

4. Discussions

4.1. Trichoderma Bio-Fertilizer Changed Aggregate Distribution and Its Associated Organic C Content

Generally, both wet and dry sieving methods are used to assess soil aggregate conditions. However, dissolved organic matter in soil would lose, and the habitat of microbes would be damaged when aggregates are sieved by the wet sieving method [13]. Therefore, it would be more convictive to assess the change of aggregate-associated organic C in soil using the dry sieving method.

Our results showed that *Trichoderma* bio-fertilizer improved the aggregation capacity of macro-aggregates (>0.25 mm), consistent with the results of Zhu et al. [9]. Wang et al. [20] also found that the proportion of 0.20–2.00 mm aggregate fraction increased, but the <0.002 mm aggregate fraction decreased after bio-fertilizer application, bio-fertilizer significantly promoted the formation of large aggregates. This might be due to the fact that amounts of fungi introduced by bio-fertilizer promoted the decomposition of soil organic matter and the formation of organo-mineral materials [2,21]. Higher soil microbial biomass after bio-fertilizer application might also promote soil aggregation [9]. Negatively charged polysaccharides and polyuronic acid, along with the growth of bacteria, were beneficial for clay bounding in soil [22], and cementation of organic matter in soil contributed to the formation of macro-aggregates. Increases of macro-aggregates protected labile C from microbial mineralization, which in turn promoted stabilization of soil aggregate [23]. Therefore, a decrease in physical protection caused by the destruction of macro-aggregates had negative effects on labile C, causing organic C mineralization [24].

Application of bio-fertilizer markedly increased soil organic C stock in bulk soil and almost each aggregate fraction, and macro-aggregate contained higher organic C than micro-aggregate. However, an opposite trend was observed by Xie et al. [15], who conducted a field experiment on Anthrosols. Diverse results among studies might be attributed to variations in binding agents during soil aggregation. Generally, increases in aggregate-associated organic C are in line with classes of aggregate fractions since organic matter is the major binding agent [25]. The organic C content was significantly higher in bio-fertilizer treatments than in CF treatment across aggregate fractions, and the BF20 treatment shared higher values. When organic material was applied, plant growth and soil microbes were both promoted [9,26], which in turn increased aggregate-associated organic C. The photosynthetic activities of algae species after bio-fertilizer application would also contribute to increasing aggregate-associated organic C [27]. Consistent with Yilmaz and Snmez [28], our results also showed that bio-fertilizer application increased organic C across aggregate fractions relative to the chemical fertilizer alone treated soil. In addition, aggregate-associated organic C was at a similar level among bio-fertilizer treatments. However, Wang et al. [20] observed that increases of organic C in macro-aggregate were higher than that in micro-aggregate after bio-fertilizer application. The diverse results suggested that soil type would be an inevitable factor in the sequestration of organic C in aggregates.

Most studies reported a linear relationship between organic C and organic materials addition rates [6,29,30]. However, in our study, organic C in aggregate fractions and bulk soil significantly increased when the bio-fertilizer application rate was below 20 t ha^{-1}, while the increase was not significant when the bio-fertilizer application rate was above it. Thus, there was a threshold effect of the bio-fertilizer application on aggregate-associated organic C in the given soil.

4.2. Mineralization of Organic C in Aggregates

It is known that the rate of organic C respiration is an effective index to assess the capability of soil C sequestration. In the present study, significantly higher CO_2-C in >2 mm aggregate was recorded. The mineralization rate of organic C in >2 mm aggregate was also significantly higher across aggregate fractions in bio-fertilizer-treated soil. These results were in line with that of Mustafa et al. [8], who recorded higher C mineralization per unit of organic C in macro-aggregate. Similarly, Kan et al. [6] found that organic C in macro-aggregate was the main source of mineralized C, of which >2 mm aggregate contributed 38.2–43.6% to the cumulative mineralization. This might be due to the fact that macro-aggregates dominated the bulk soil (>2 mm aggregate occupies 34.90–48.94% and 0.25–2 mm aggregate occupies 26.08–36.60% in the present study). Organic C in macro-aggregate, mainly originated from plant materials, was more labile. Additionally, larger pores in macro-aggregate increased the transposition of oxygen and microbial activities, which promoted the process of organic C mineralization in macro-aggregate. In the case of Anthrosols, Xie et al. [15] found greater CO_2-C produced from micro-aggregate than macro-aggregate, suggesting that mineralizable C in micro-aggregate was much higher than that in macro-aggregate. However, Rabbi et al. [31] showed that there was no difference in organic C mineralization between macro-aggregate and micro-aggregate in Oxisols. Thus, mineralization of aggregate-associated organic C might be a better reflection of C sequestrated in aggregates. Generally, organic C in micro-aggregate was relatively stable for soil in which organic C dominated soil aggregation [32].

When compared with CF, the cumulative mineralization of organic C in aggregates and bulk soil significantly increased with increasing bio-fertilizer application rates. This might be due to the fact that bio-fertilizer application enhanced amounts of available nutrients and promoted microbial activities, which both contributed to the mineralization of organic C [33,34]. Similarly, Xie et al. [15] showed that manure application significantly increased the mineralization of aggregate-associated organic C, and that proportions of organic C mineralized were also similar in the surface layer. Notably, BF20 treatment shared a significantly lower rate of organic C respiration across aggregate fractions (except for <0.053 mm aggregate). The result might indicate that the bio-fertilizer application rate was an important factor in regulating the mineralization of aggregate-associated organic C. The decreased C mineralization in bio-fertilizer application treatments, especially in BF20, might be explained by the dilution effect. Namely, the significant increase of organic C mineralization was diluted by the large increase of aggregate-associated organic C. Generally, organic C in aggregate fractions and bulk soil in the BF20 treatment was more resistant to mineralization, which would be beneficial for soil C sequestration.

4.3. Bio-Fertilizer Alter Microbial Community in Aggregates

PLFA can well reflect changes in soil microbial communities influenced by fertilization. Our study showed that bio-fertilizer significantly increased total PLFA and each microbial group PLFA across aggregates and in bulk soil. Amounts of organic material introduced into the soil would provide extra nutrients and energy to soil microbes, which might directly increase the PLFAs of soil microorganisms within aggregates [35]. This was confirmed by the significant positive correlations between PLFAs and aggregate-associated organic C. In line with the results of Jiang et al. [36], our results also showed that total PLFA, bacteria PLFA and fungi PLFA in macro-aggregate were greater than that in micro-aggregate. It might be that levels of organic C and soil microbes were relatively low in micro-aggregates compared to those in macro-aggregates and that the turnover rate of C was slow. Differently, Wang et al. [37] showed that PLFAs of bacteria and fungi in macro-aggregate were significantly increased by manure application while no significant difference was observed in micro-aggregate. Due to the heterogeneity of soil properties, differences in microbial community composition across aggregate fractions would appear. Aggregates with larger particle sizes supported higher nutrient levels, which were conducive to microbial colonization [2,38].

Generally, bio-fertilizer high in C/N ratio would be more beneficial to fungal growth [39], which led to increased fungi PLFA/bacteria PLFA. In addition, significant changes in bacterial PLFA and fungal PLFA represented variations in soil microbial community, though bacteria were the dominant group among different treatments. It was believed that the higher ratio of fungi PLFA to bacteria PLFA indicated a more stable soil ecosystem and better soil quality [40]. The higher fungi PLFA/bacteria PLFA in BF20 treatment suggested that appropriate bio-fertilizer application helped maintain the stability of the soil ecosystem. Our results showed that bio-fertilizer significantly increased contents of >0.25 mm aggregates and its fungi PLFA, suggesting that fungi contributed to the formation of macro-aggregate and soil aggregate stability via filamentous growth and excretion production [41]. Thus, bio-fertilizer could increase the content of macro-aggregate by promoting fungal growth, and fungi were important factors for bio-fertilizer to promote the formation of macro-aggregates. Additionally, fungi could decompose foreign nutrients through hyphal movement and had high assimilation efficiency of C source, while bacteria did not have this advantage. However, extra organic material usually promoted the growth of G− bacteria, which preferred plant-derived carbon sources, while the G+ community usually participated in organic matter and litter decomposition [42]. A lower G+ PLFA/G− PLFA meant a better soil nutritional condition [43]. Thus, decreased G+ PLFA/G− PLFA after bio-fertilizer application indicated that the soil environment shifted to more eutrophic conditions. In addition, the fact that the ratio of G+ PLFA/G− PLFA was negatively correlated with aggregate-associated organic C also indicated the improved soil quality after bio-fertilizer application. Therefore, the bio-fertilizer application was beneficial to improving soil nutrient status and ecological buffer capacity of large aggregates in the farmland ecosystem. Though the mineralization of aggregate-associated organic C was positively correlated with fungi, *Trichoderma* was not distinguished from fungi in the present study. The abundance of specific microorganisms was also not assessed, which played vital roles in organic carbon turnover. Thus, further studies focusing on the effects of bio-fertilizer on soil microbial community composition are needed, which may help to develop an environmental bio-fertilizer application pattern in agricultural production.

5. Conclusions

Bio-fertilizer considerably increased proportions of macro-aggregate, organic C sequestration and microbial PLFAs across aggregate fractions and bulk soil. The increase of MWD in BF20 reached 35.6%, significantly higher than other treatments. Bio-fertilizer increased the fungi PLFA/bacteria PLFA but decreased G+ PLFA/G− PLFA, and the BF20 shared the greatest changes. Increases of aggregate-associated organic C of >2, 0.25–2, 0.053–0.25 and <0.053 mm in BF20 were 18.15%, 18.60%, 17.75% and 18.46%, respectively, relative to CF. However, the cumulative mineralization of organic C was relatively low in BF20. Thus, the promotion of organic C stock was generally higher in BF20, while the proportion of organic C mineralization was relatively low across aggregate fractions. Correlation analysis showed that microbial community in aggregate was correlated with increases of soil C, of which bacteria and fungi contributed more than actinomycetes. This study highlighted the vital role of *Trichoderma* bio-fertilizer in regulating C mineralization at the aggregate level and provided scientific bases for bio-fertilizer application in Shajiang Calci-Aquic Vertosol. However, we did not determine the distinct keystone taxa and their co-occurrence patterns. Further study concerning keystone taxa at the aggregate level after bio-fertilizer application is needed.

Author Contributions: Writing—original draft preparation, L.Z. and Y.C.; writing—review and editing, M.C., C.S. and T.L.; visualization, Y.Z.; funding acquisition, L.L. All authors have read and agreed to the published version of the manuscript.

Funding: This research was financially supported by the Scientific and Technological Project of Henan Provincial Science and Technology Department of China (222102320276), the National Key Research and Development Program of China (2018YFD0300704).

Institutional Review Board Statement: Not applicable.

Informed Consent Statement: Not applicable.

Data Availability Statement: The data presented in this study are available upon request from the corresponding author.

Conflicts of Interest: The authors declare no conflict of interest.

References

1. Bardgett, R.D.; McAlister, E. The measurement of soil fungal: Bacterial biomass ratios as an indicator of ecosystem self-regulation in temperate meadow grasslands. *Biol. Fert. Soils* **1999**, *29*, 282–290. [CrossRef]
2. Zhang, H.J.; Wang, S.J.; Zhang, J.X.; Tian, C.J.; Luo, S.S. Biochar application enhances microbial interactions in mega-aggregates of farmland black soil. *Soil Till. Res.* **2021**, *213*, 105145. [CrossRef]

3. Mustafa, A.; Xu, M.G.; Ali Shah, S.A.; Abrar, M.M.; Sun, N.; Wang, B.R.; Cai, Z.J.; Saeed, Q.; Naveed, M.; Mehmood, K.; et al. Soil aggregation and soil aggregate stability regulate organic carbon and nitrogen storage in a red soil of southern China. *J. Environ. Manag.* **2020**, *270*, 110894. [CrossRef] [PubMed]

4. Liao, H.; Zhang, Y.C.; Zuo, Q.Y.; Du, B.B.; Chen, W.L.; Wei, D.; Huang, Q.Y. Contrasting responses of bacterial and fungal communities to aggregate-size fractions and longterm fertilizations in soils of northeastern China. *Sci. Total Environ.* **2018**, *635*, 784–792. [CrossRef] [PubMed]

5. Jiang, Y.J.; Qian, H.Y.; Wang, X.Y.; Chen, L.J.; Liu, M.Q.; Li, H.X.; Sun, B. Nematodes and microbial community affect the sizes and turnover rates of organic carbon pools in soil aggregates. *Soil Biol. Biochem.* **2018**, *119*, 22–31. [CrossRef]

6. Kan, Z.R.; Ma, S.T.; Liu, Q.Y.; Liu, B.Y.; Virk, A.L.; Qi, J.Y.; Zhao, X.; Rattan, L.; Zhang, H.L. Carbon sequestration and mineralization in soil aggregates under long-term conservation tillage in the North China Plain. *Catena* **2020**, *188*, 104428. [CrossRef]

7. Li, X.P.; Liu, C.L.; Zhao, H.; Gao, F.; Ji, G.N.; Hu, F.; Li, H.X. Similar positive effects of beneficial bacteria, nematodes and earthworms on soil quality and productivity. *Appl. Soil. Ecol.* **2018**, *130*, 202–208. [CrossRef]

8. Mustafa, A.; Xu, H.; Ali Shah, S.A.; Abrar, M.M.; Maitlo, A.A.; Kubar, K.A.; Saeed, Q.; Kamran, M.; Naveed, M.; Wang, B.R.; et al. Long-term fertilization alters chemical composition and stability of aggregate-associated organic carbon in a Chinese red soil: Evidence from aggregate fractionation, C mineralization, and ^{13}C NMR analyses. *J. Soil Sediment.* **2021**, *21*, 2483–2496. [CrossRef]

9. Zhu, L.X.; Zhang, F.L.; Li, L.L.; Liu, T.X. Soil C and aggregate stability were promoted by bio-fertilizer on the North China Plain. *J. Soil Sci. Plant Nutr.* **2021**, *21*, 2355–2363. [CrossRef]

10. Li, Y.P.; Wang, J.; Shao, M.A. Application of earthworm cast improves soil aggregation and aggregate-associated carbon stability in typical soils from Loess Plateau. *J. Environ. Manag.* **2021**, *278*, 111504. [CrossRef]

11. Qiu, L.P.; Zhu, H.S.; Liu, J.; Yao, Y.F.; Wang, X.; Rong, G.H.; Zhao, X.N.; Shao, M.A.; Wei, X.R. Soil erosion significantly reduces organic carbon and nitrogen mineralization in a simulated experiment. *Agr. Ecosyst. Environ.* **2021**, *307*, 107232. [CrossRef]

12. Reeves, S.H.; Somasundaram, J.; Wang, W.J.; Heenan, M.A.; Finn, D.; Dalal, R.C. Effect of soil aggregate size and long-term contrasting tillage, stubble and nitrogen management regimes on CO_2 fluxes from a Vertisol. *Geoderma* **2019**, *337*, 1086–1096. [CrossRef]

13. Wang, X.Y.; Bian, Q.; Jiang, Y.J.; Zhu, L.Y.; Chen, Y.; Liang, Y.T.; Sun, B. Organic amendments drive shifts in microbial community structure and keystone taxa which increase C mineralization across aggregate size classes. *Soil Biol. Biochem.* **2021**, *153*, 108062. [CrossRef]

14. Rabbi, S.M.F.; Wilson, B.R.; Lockwood, P.V.; Daniel, H.; Young, I.M. Soil organic carbon mineralization rates in aggregates under contrasting land uses. *Geoderma* **2014**, *216*, 10–18. [CrossRef]

15. Xie, J.Y.; Hou, M.M.; Zhou, Y.T.; Wang, R.J.; Zhang, S.L.; Yang, X.Y.; Sun, B.H. Carbon sequestration and mineralization of aggregate-associated carbon in an intensively cultivated Anthrosol in north China as affected by long term fertilization. *Geoderma* **2017**, *296*, 1–9. [CrossRef]

16. Bai, N.; Zhang, H.; Li, S.; Zheng, X.; Zhang, J.; Zhang, H.; Zhou, S.; Sun, H.; Lv, W. Long-term effects of straw and straw-derived biochar on soil aggregation and fungal community in a rice–wheat rotation system. *PeerJ* **2019**, *6*, e6171. [CrossRef] [PubMed]

17. Yang, X.Y.; Sun, B.H.; Zhang, S.L. Trends of yield and soil fertility in a long-term wheat-maize System. *J. Integr. Agric.* **2014**, *13*, 402–414. [CrossRef]

18. Bligh, E.G.; Dyer, W.J. A rapid method of total lipid extraction and purification. *Can. J. Biochem. Physiol.* **1959**, *37*, 911–917. [CrossRef]

19. Luo, S.S.; Wang, S.J.; Tian, L.; Shi, S.H.; Xu, S.Q.; Yang, F.; Li, X.J.; Wang, Z.C.; Tian, C.J. Aggregate-related changes in soil microbial communities under different ameliorant applications in saline-sodic soils. *Geoderma* **2018**, *329*, 108–117. [CrossRef]

20. Wang, M.; Chen, S.B.; Han, Y.; Chen, L.; Wang, D. Responses of soil aggregates and bacterial communities to soil-Pb immobilization induced by bio-fertilizer. *Chemosphere* **2019**, *220*, 828–836. [CrossRef]

21. Kong, A.Y.Y.; Scow, K.M.; Córdova-Kreylos, A.L.; Holmes, W.E.; Six, J. Microbial community composition and carbon cycling within soil microenvironments of conventional, low-input, and organic cropping systems. *Soil Biol. Biochem.* **2011**, *43*, 20–30. [CrossRef]

22. Guibaud, G.; Bordas, F.; Saaid, A.; Paul, D.; Van Hullebusch, E. Effect of pH on cadmium and lead binding by extracellular polymeric substances (EPS) extracted from environmental bacterial strains. *Colloid. Surf. B* **2008**, *63*, 48–54. [CrossRef]

23. Six, J.; Conant, R.T.; Paul, E.A.; Paustian, K. Stabilization mechanisms of soil organic matter: Implications for C-saturation of soils. *Plant Soil* **2002**, *241*, 155–176. [CrossRef]

24. Somasundaram, J.; Chaudhary, R.; Kumar, D.A.; Biswas, A.K.; Sinha, N.K.; Mohanty, M.; Hati, K.M.; Jha, P.; Sankar, M.; Patra, A.K.; et al. Effect of contrasting tillage and cropping systems on soil aggregation, carbon pools and aggregate-associated carbon in rainfed Vertisols. *Eur. J. Soil Sci.* **2018**, *69*, 879–891. [CrossRef]

25. Six, J.; Bossuyt, H.; Degryze, S.; Denef, K. A history of research on the link between (micro)aggregates, soil biota, and soil organic matter dynamics. *Soil Till. Res.* **2004**, *79*, 7–31. [CrossRef]

26. Bei, S.H.; Li, X.; Kuyper, T.W.; Chadwick, D.R.; Zhang, J.L. Nitrogen availability mediates the priming effect of soil organic matter by preferentially altering the straw carbon-assimilating microbial community. *Sci. Total Environ.* **2022**, *815*, 152882. [CrossRef]

27. Trejo, A.; de-Bashan, L.E.; Hartmann, A.; Hernandez, J.P.; Rothballer, M.; Schmid, M.; Bashan, Y. Recycling waste debris of immobilized microalgae and plant growth-promoting bacteria from wastewater treatment as a resource to improve fertility of eroded desert soil. *Environ. Exp. Bot.* **2012**, *75*, 65–73. [CrossRef]

28. Yilmaz, E.; Snmez, M. The role of organic/bio–fertilizer amendment on aggregate stability and organic carbon content in different aggregate scales. *Soil Till. Res.* **2017**, *168*, 118–124. [CrossRef]

29. Du, Z.L.; Zhao, J.K.; Wang, Y.D.; Zhang, Q.Z. Biochar addition drives soil aggregation and carbon sequestration in aggregate fractions from an intensive agricultural system. *J. Soil Sediment.* **2017**, *17*, 581–589. [CrossRef]

30. Shao, H.Y.; Li, Z.Y.; Liu, D.; Li, Y.F.; Lu, L.; Wang, X.D.; Zhang, A.F.; Wang, Y.L. Effects of manure application rates on the soil carbon fractions and aggregate stability. *Environ. Sci.* **2019**, *40*, 4691–4699. [CrossRef]

31. Rabbi, S.M.F.; Wilson, B.R.; Lockwood, P.V.; Daniel, H.; Young, I.M. Aggregate hierarchy and carbon mineralization in two Oxisols of New South Wales, Australia. *Soil Till. Res.* **2015**, *146*, 193–203. [CrossRef]

32. Bronick, C.J.; Lal, R. Soil structure and management: A review. *Geoderma* **2005**, *124*, 3–22. [CrossRef]

33. Chen, X.F.; Liu, M.; Jiang, C.Y.; Wu, M.; Li, Z.P. Organic carbon mineralization in aggregate fractions of red paddy soil under different fertilization treatments. *Sci. Agric. Sin.* **2018**, *51*, 3325–3334. (In Chinese) [CrossRef]

34. Ashraf, M.N.; Hu, C.; Wu, L.; Duan, Y.; Zhang, W.; Aziz, T.; Cai, A.; Abrar, M.M.; Xu, M. Soil and microbial biomass stoichiometry regulate soil organic carbon and nitrogen mineralization in rice-wheat rotation subjected to long-term fertilization. *J. Soil Sediment.* **2020**, *20*, 3103–3113. [CrossRef]

35. Marks, E.A.N.; Miñón, J.; Pascual, A.; Montero, O.; Navas, L.M.; Rad, C. Application of a microalgal slurry to soil stimulates heterotrophic activity and promotes bacterial growth. *Sci. Total Environ.* **2017**, *605–606*, 610–617. [CrossRef]

36. Jiang, Y.J.; Sun, B.; Jin, C.; Wang, F. Soil aggregate stratification of nematodes and microbial communities affects the metabolic quotient in an acid soil. *Soil Biol. Biochem.* **2013**, *60*, 1–9. [CrossRef]

37. Wang, Y.; Hu, N.; Ge, T.; Kuzyakov, Y.; Wang, Z.; Li, Z.; Tang, Z.; Chen, Y.; Wu, C.; Lou, Y. Soil aggregation regulates distributions of carbon, microbial community and enzyme activities after 23-year manure amendment. *Appl. Soil Ecol.* **2017**, *111*, 65–72. [CrossRef]

38. Liao, H.; Zhang, Y.; Wang, K.; Hao, X.; Chen, W.; Huang, Q. Complexity of bacterial and fungal network increases with soil aggregate size in an agricultural Inceptisol. *Appl. Soil Ecol.* **2020**, *154*, 103640. [CrossRef]

39. Wang, Q.Q.; Liu, L.L.; Li, Y.; Song, Q.; Wang, C.J.; Cai, A.D.; Wu, L.; Xu, M.G.; Zhang, W.J. Long-term fertilization leads to specific PLFA finger-prints in Chinese Hapludults soil. *J. Integr. Agric.* **2020**, *19*, 1354–1362. [CrossRef]

40. Vries, F.T.D.; Hoffland, E.; Eekeren, N.V.; Brussaard, L.; Bloem, J. Fungal/bacterial ratios in grasslands with contrasting nitrogen management. *Soil Biol. Biochem.* **2006**, *38*, 2092–2103. [CrossRef]

41. Lehmann, A.; Rillig, M.C. Understanding mechanisms of soil biota involvement in soil aggregation: A way forward with saprobic fungi? *Soil Biol. Biochem.* **2015**, *88*, 298–302. [CrossRef]

42. Naylor, D.; De Graaf, S.; Purdom, E.; Coleman-Derr, D. Drought and host selection influence bacterial community dynamics in the grass root microbiome. *ISME J.* **2017**, *11*, 2691. [CrossRef]

43. Bhattacharyya, R.; Bhatia, A.; Das, T.K.; Lata, S.; Kumar, A.; Tomer, R.; Singh, G.; Kumar, S.; Biswas, A.K. Aggregate-associated N and global warming potential of conservation agriculture-based cropping of maize-wheat system in the north-western Indo-Gangetic Plains. *Soil Till. Res.* **2018**, *182*, 66–77. [CrossRef]

 *agriculture*

Article

Carbon Storage Potential of Agroforestry System near Brick Kilns in Irrigated Agro-Ecosystem

Nayab Komal [1], Qamar uz Zaman [1,*], Ghulam Yasin [2], Saba Nazir [1], Kamran Ashraf [3], Muhammad Waqas [1], Mubeen Ahmad [1], Ammara Batool [1], Imran Talib [1] and Yinglong Chen [4,5,*]

1 Department of Environmental Sciences, The University of Lahore, Lahore 54590, Pakistan; nayabkomal5@gmail.com (N.K.); sabawarraich989@gmail.com (S.N.); wikash.grt@gmail.com (M.W.); mubeeen741@gmail.com (M.A.); ammarabatool148@gmail.com (A.B.); imran.talib@envs.uol.edu.pk (I.T.)
2 Department of Forestry, Range and Wildlife Management, Baghdad Ul Jadeed Campus, The Islamia University, Bahawalpur 63100, Pakistan; yasin_2486@yahoo.com
3 Department of Food Sciences, Government College University Faisalabad Sahiwal Campus, Sahiwal 57000, Pakistan; kamran2417@gmail.com
4 The UWA Institute of Agriculture, School of Agriculture and Environment, The University of Western Australia, Perth, WA 6009, Australia
5 Institute of Soil and Water Conservation, Northwest A&F University, Yangling 712100, China
* Correspondence: qamar.zaman1@envs.uol.edu.pk (Q.u.Z.); yinglong.chen@uwa.edu.au (Y.C.)

Abstract: The current study was conducted to estimate the carbon (C) storage status of agroforestry systems, via a non-destructive strategy. A total of 75 plots (0.405 ha each) were selected by adopting a lottery method of random sampling for C stock estimations for soil, trees and crops in the Mandi-Bahauddin district, Punjab, Pakistan. Results revealed that the existing number of trees in selected farm plots varied from 25 to 30 trees/ha. Total mean tree carbon stock ranged from 9.97 to 133 Mg C ha^{-1}, between 5–10 km away from the brick kilns in the study area. The decreasing order in terms of carbon storage potential of trees was *Eucalyptus camaldulensis* > *Syzygium cumin* > *Popolus ciliata* > *Acacia nilotica* > *Ziziphus manritiana* > *Citrus sinensis* > *Azadirachtta Indica* > *Delbergia sisso* > *Bambusa vulgaris* > *Melia azadarach* > *Morus alba*. Average soil carbon pools ranged from 10.3–12.5 Mg C ha^{-1} in the study area. Meanwhile, maximum C stock for wheat (2.08×10^6 Mg C) and rice (1.97×10^6 Mg C) was recorded in the cultivated area of Tehsil Mandi-Bahauddin. The entire ecosystem of the study area had an estimated woody vegetation carbon stock of 68.5 Mg C ha^{-1} and a soil carbon stock of 10.7 Mg C ha^{-1}. These results highlight that climate-smart agriculture has great potential to lock up more carbon and help in the reduction of CO_2 emissions to the atmosphere, and can be further used in planning policies for executing tree planting agendas on cultivated lands and for planning future carbon sequestration ventures in Pakistan.

Keywords: agroforestry; brick kilns; carbon emissions; climate change; carbon sinks; carbon stock

Citation: Komal, N.; Zaman, Q.u.; Yasin, G.; Nazir, S.; Ashraf, K.; Waqas, M.; Ahmad, M.; Batool, A.; Talib, I.; Chen, Y. Carbon Storage Potential of Agroforestry System near Brick Kilns in Irrigated Agro-Ecosystem. *Agriculture* **2022**, *12*, 295. https://doi.org/10.3390/agriculture12020295

Academic Editor: Rosa Francaviglia

Received: 18 January 2022
Accepted: 16 February 2022
Published: 18 February 2022

1. Introduction

Pakistan is predicted to be among the ten countries most affected by climate change, according to the 2019 Global Climate Risk Index [1]. Global climate change, by increasing the amount of greenhouse gases (GHG) in the atmosphere, is causing severe environmental and climatic effects. Carbon dioxide (CO_2) is one of the most commonly highlighted greenhouse gasses. Global climate change and the increase in the trend of CO_2 emissions are a growing concern today [2]. Pakistan is confronting this powerful danger to social, environmental, and economic development [3]. The impacts of climate change can be categorized into extreme and non-extreme types [4,5]. The World Bank [6] recognizes five foremost factors through which climate change will affect agricultural production: change in precipitation pattern and temperature, climatic variability, CO_2 fertilization and surface water runoff. Reilly et al. [7] found that higher rainfall results in reduction of yield.

Variations in climate cause many people to move into poverty and food insecurity [8]. In Pakistan, the general origin of air pollution is customarily untidy industrial buildings of brick formation, present in the peri-metropolitan and rural regions [9]. Inferior quality fuels, including corncobs, rubber tires, rice straw, bagasse, rice husk, coal, oil and wood, used in these brick kilns produce fly ash particles that deposit on nearby plants affecting their photosynthetic potential [6]. Brick kilns release over 1072 million tons of CO_2 emissions into the atmosphere per year, making 2.7% of total emissions. Most of the brick kilns in the rural areas in Pakistan use conventional technology that is very dangerous from an environmental aspect. According to the Punjab Disaster Management Authority (PDMA), 37.4% of brick kilns have moved to zig-zag technology in Punjab alone [1]. All the types of fuel utilized in this type of kiln cause a high concentration of pollutants in gaseous form in the air, with destructive effects on the atmosphere, plants and people [10,11]. Islam et al. [12] estimated that the soil close to the brick kiln was reduced in quantity when compared with the same soil further from brick kiln, showing a variation in agricultural production. The mean values of total nitrogen, available phosphorus and sulfur were significantly less in the soil samples close to the brick kiln (0.05%, 12.4, and 8.36 ppm, respectively) than those in the soil further from the brick kiln (0.06%, 24.6, and 11.7 ppm, respectively).

Agriculture is a key economic sector that contributes 21% to the gross domestic product (GDP), employs 45% of the total workforce and contributes about 60% to exports [13]. Changing climatic variables, particularly temperature and rainfall, will introduce several challenges to agriculture in the future. Changes in the frequency and intensity of droughts, flooding, and storm damage are anticipated [13]. Agroforestry could provide adaptation to this climate change [14] by protecting crops from temperature elevation. On the other hand, this causes a decrease in soil evaporation, wind seeds, and transpiration of crops. The carbon (C) sequestration above ground could easily be increased by planting trees and this also increases carbon in the soil on the land where crops are cultivated [15,16]. The general sequestration of carbon due to such actions has been assumed to be 9, 21, 50, and 63 Mg C ha^{-1} in temperate, sub-humid, semiarid, and humid regions, respectively [17]. The planting of trees together with crops has many advantages, involving higher soil richness, limitation of soil erosion, lower water logging, decreased fermentation and eutrophication of streams and rivers, enhancement of local biodiversity, and reduction of pressure on common forests for fuel [18–20]. At a global scale, unproductive croplands of about 630 million ha could be used for agroforestry as part of an ecological engineering tool to potentially sequester 586,000 Mg C year^{-1} by 2040. Moreover, in present national and global monitoring protocols for carbon, there is a need to include agroforestry in C stocks to estimate the share of this abandoned pool in a precise way. Agroforestry systems are considered as a carbon sink to sequester CO_2, so there is a need to evaluate the carbon sequestration potential of agroforestry systems in order to decrease emissions near brick kilns on irrigated land. The main objectives of the current investigation were (1) to assess and quantify the potential of agroforestry systems in C stock at various distances from brick kilns in an irrigated agro-ecosystem, (2) to discover differences between agroforestry systems and the carbon capturing capacity of crops, trees and soil organic carbon (SOC).

2. Materials and Methods

2.1. Study Locations and Sampling Methodology

The present study was conducted in the Mandi-Bahauddin district, central Punjab, Pakistan (Figure 1). According to the population survey in 2017, Mandi-Bahauddin is the 41st city by population. This city is located between the river Chenab (39 km in south) and the river Jhelum (12 km in north). It lies between $30°8'$ to $32°40'$ north latitude, and $73°36'$ to $73°37'$ east longitude. It has a total area of 2673 km^2. Major rivers are River Jhelum, with Rasool Barrage & Chenab and the Qadirabad Barrage. According to the Pakistan Bureau of Statistics (2019), the total population of Mandi-Bahuddin is 198,609 with a population density of 30 km^2. The annual change in population is 1.68%. The elevation of this district is 220 m above the sea level. The weather conditions experience a maximum average temperature of 45 °C and minimum average temperature of 3 °C, with average rainfall of 388 mm and wind direction mostly from north to east (Pakistan meteorological department, PMD). According to the US Department of Agriculture classification system, the soil type of the study area falls in the category of loam to clay loam soil with organic matter ranging from 0.50–1.01%. Most farmers in these areas use conventional practices (chemical fertilizer) for crop production instead of using organic methods for crop production to improve soil health. According to the agricultural department, the main crops in study area are wheat, rice and sugarcane and the total cultivated area is 214,348.83 ha. Maize and pulses are also grown in smaller areas.

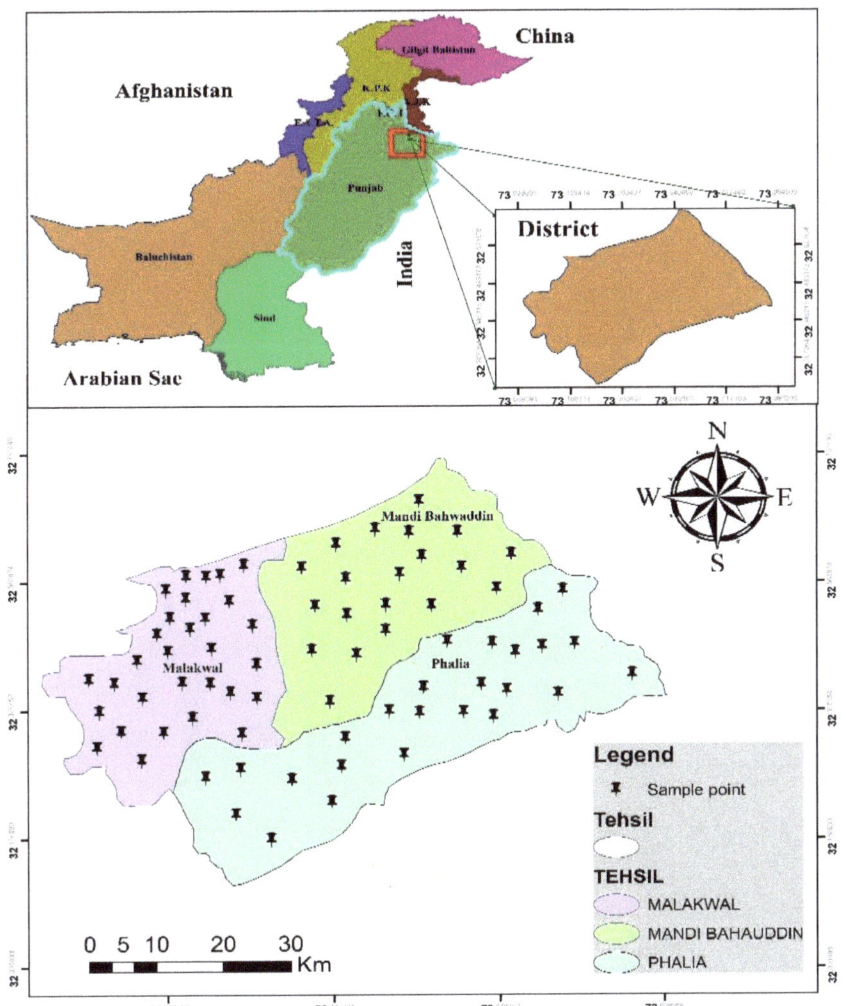

Figure 1. Map of the study area and sampling points.

2.2. Above and Below Ground Trees' Biomass Carbon Estimation

Field visits were carried out from November 2020 to May 2021 at regular intervals for the collection of data in 3 selected Tehsils, i.e., Mandi-Bahauddin, Phalia, Malakwal. A total of 75 quadrate plots of 0.405 ha = 1 acre each with agroforestry practices were selected and a tree inventory such as diameter at breast height in cm (DBH) and height (m) was taken for each plot (Table 1). By adopting a lottery method of random sampling, the sample collection sites were selected on the basis of brick kiln location at 0, 5 and 10 km distance. Mean values of tree height and diameter at breast height of potential tree species at various distances from the brick kilns are depicted in Table 2.

Table 1. Number of plots measured from selected Tehsils in Mandi-Bahauddin sampling.

District	Tehsil	No. of Measured Plots (0.405 ha)
Mandi-Bahauddin	Mandi-Bahauddin	30
	Phalia	25
	Malakwal	20

Table 2. Diameter at Breast Height (DBH) and Height of Potential Tree Species at Various Distance from Brick Kilns in the Agroforestry System of Mandi-Bahauddin district (Mean Values).

Tehsils	Potential Tress Species	Near Brick Kilns		5 km from Brick Kilns		10 km from Brick Kilns	
		Height (m)	DBH (cm)	Height (m)	DBH (cm)	Height (m)	DBH (cm)
Mandi-Bahauddin	A. nilotica	8.8	6.5	7.7	8.3	10.3	8.6
	A. indica	7.7	6.5	7.8	9.6	9.1	8.6
	M. azadarach	8.2	6.5	7.9	9.0	9.5	8.5
	C. sinensis	6.8	6.4	7.8	8.6	8.2	8.6
	D. sisso	7.3	6.6	7.8	8.7	8.7	8.8
	E. camaldulensis	6.6	6.6	8.1	8.4	8.0	8.8
	P. ciliata	7.5	6.7	7.8	8.3	8.9	8.8
	S. cumin	7.8	6.6	7.9	9.0	9.1	8.7
	Z. manritiana	8.2	6.5	7.9	9.2	9.1	8.9
	M. alba	7.8	6.5	7.8	9.4	9.2	8.7
	B. vulgaris	8.5	6.6	7.6	9.2	9.9	8.9
Phalia	A. nilotica	7.9	6.3	7.7	7.4	6.7	5.8
	A. indica	8.1	6.4	7.5	3.0	7.6	5.2
	M. azadarach	8.5	6.5	7.7	8.6	6.4	6.3
	C. sinensis	8.3	6.3	7.6	8.3	7.3	6.5
	D. sisso	8.1	6.4	7.7	3.2	7.6	6.4
	E. camaldulensis	7.3	6.4	7.6	3.3	8.3	6.5
	P. ciliata	6.3	6.4	7.7	3.1	7.0	6.6
	S. cumin	8.0	6.4	7.6	8.7	8.1	6.6
	Z. manritiana	8.4	6.4	7.4	8.1	8.0	6.5
	M. alba	8.0	6.4	7.6	8.3	8.0	6.4
	B. vulgaris	8.2	6.4	7.2	8.1	8.6	6.3
Malakwal	A. nilotica	10.9	6.0	7.3	3.2	11.1	5.7
	A. indica	10.2	5.3	7.3	3.5	7.3	3.2
	M. azadarach	10.5	5.3	6.6	10.6	7.0	3.2
	C. sinensis	3.5	5.3	7.3	11.3	8.8	6.7
	D. sisso	3.7	6.0	7.4	11.4	8.3	3.3
	E. camaldulensis	3.1	6.1	7.2	11.5	8.5	7.3
	P. ciliata	3.3	6.1	7.2	3.6	7.3	3.3
	S. cumin	3.5	6.0	7.3	3.6	7.0	3.2
	Z. manritiana	3.7	5.3	6.7	11.8	7.1	3.3
	M. alba	3.8	6.0	5.3	11.3	6.8	3.2
	B. vulgaris	3.4	6.1	5.4	10.6	6.6	8.1

Acacia nilotica [Fabaceae], *Delbergia sisso* [Fabaceae], *Melia azedarach* [Meliaceae], *Citrus reticulate* [Rutaceae], *Popolus ciliate* [Salicaceae], *Eucalyptus camaldulensis* [Myrtaceae], *Melia azedarach* [Meliaceae], *Populus deltoides* [Salicaceae], *Syzygium cumini* [Myrtaceae], *Ziziphus mauritiana* [Rhamnaceae], *Azadirachtta indica* [Meliaceae], *Morus alba* [Moraceae] and *Bambusa vulgaris* [Poaceae] were the most commonly planted tree species in and along the farm fields in all three tehsils.

2.3. Total Carbon Stock Estimate

Allometric equations from the literature were used for the estimation of tree biomass and where appropriate corrected for log bias (Table 3). In case of non-availability of allometric equations, 26% of the above ground biomass was assumed as below ground biomass [21–23]. Next, biomass of individual trees was scaled to biomass/plot, biomass/hectare, and carbon stock/hectare. Contents of carbon were measured from biomass by presumption that the dry mass contains 48.1% of carbon [21,24]. The total tehsil tree carbon was estimated by multiplying carbon amount per hectare from sampled plots by the total area of the tehsil.

Table 3. Allometric equations for the calculation of above and below ground biomass.

Species	Component	Allometeric Equations	Source	R^2	MSE
A. nilotica	AGB	$LogY = -1.0646 + 0.9098 \times logD^2H$	[25]	0.96	-
	BGB	$LogY = -1.3952 + 0.8253 \times logD^2H$	[25]	0.92	-
A. indica	AGB	$LnY = -3.1114 + 0.9719 \times \ln D^2H$	[26]	0.97	0.116
	BGB	$BGB = AGB \times 0.26$	[27]	-	-
E. camaldulensis	AGB	$LnY = -2.2660 + 2.4663 \times \ln D^2H$	[28]	0.99	-
	BGB	$BGB = AGB \times 0.26$	[27]	-	-
M. azedarach	AGB	$Y = 42.321 + 9.52 \times 10^{-5} \times D^2H$	[29]	0.74	-
	BGB	$BGB = AGB \times 0.26$	[27]	-	-
M. alba	AGB	$LnY = -3.1114 + 0.9719 \times \ln D^2H$	[26]	0.97	0.116
	BGB	$BGB = AGB \times 0.26$	[27]	-	-
P. deltoides	AGB	$Y = 173.144 \times [1 + (2.956 - B \times 0.120 \times DBH)]^{-1}$	[30]	0.99	-
	BGB	$Y = 69.105 \times [1 + (3.273 - 0.077 \times DBH)]^{-1}$	[30]	0.98	-
S. cumini	AGB	$LogY = -1.2066 + 0.9872 \times logD^2H$	[31]	0.97	-
	BGB	$BGB = AGB \times 0.26$	[27]	-	-
Z. mauritiana	AGB	$LnY = -3.1114 + 0.9719 \times \ln D^2H$	[26]	0.97	0.116
	BGB	$BGB = AGB \times 0.26$	[27]	-	-
D. sissoo	Bole	Tree age < 4 $Y = -0.367 + 1.3457 \times DBH$	[32]	0.97	-
	Branch	$Y = -1.4581 + 0.7708 \times DBH$	-	0.94	-
	Twig	$Y = -0.2932 + 0.1461 \times DBH$	-	0.94	-
	Leaf	$Y = -0.4501 + 0.283 \times DBH$	-	0.94	-

AGB = aboveground biomass; B = belowground biomass; D = tree diameter at 1.3 m (cm); H = total tree height (m), BA = individual tree basal area (cm^2).

2.4. Soil Sampling Collection

From 0, 5 and 10 km away from the brick kilns, soil was sampled in a random subset of plots to represent the major tree and crop combinations. Soil samples (n = 420) were collected at a depth of 0–30 cm near the base of a randomly selected tree, from the four cardinal directions. Samples were stored in polythene bags and analyzed at the Soil and Water Testing Laboratory for Research, Bahawalpur. A 100 cm^3 stainless-steel cylinder was used to measure soil bulk density. After being air-dried and passed through a 2 mm sieve, organic carbon was measured using the Walkley–Black method [33]. To calculate the soil carbon per hectare, the values of bulk density, soil depth, and percentage of organic carbon were then multiplied [34].

2.5. Crop Carbon Stock Determination

Wheat and rice plants were manually harvested to the required depth, sun dried and then weighed with a spring balance to measure above and below ground biomass per plot, converted into Mg ha^{-1}. From each plot at different locations, a different number of tillers of wheat and rice crops were selected randomly from an area of 1 m^2. Above ground and below ground C stock in wheat and rice samples was determined by multiplying the respective above ground and below ground biomass with carbon conversion factor of 0.45, as explained by Prommer et al. [35]. Next, biomass of individual crop was scaled to biomass/hectare, carbon stock/hectare, and finally on a tehsil basis.

2.6. Data Analysis

Collected data was analyzed using "Statistix 8.1" and "Statistix 10" statistical packages. Data regarding analytical analysis of AGB, BGB and SOC were analyzed using descriptive statistics. Graphical work was performed using Microsoft Office software (Version, 2016; Microsoft Corporation, Albuquerque, NM, USA).

3. Results

3.1. Tree Abundance and Distribution

Variations in trees' abundance and distribution were observed in the Mandi-Bahauddin district. All the tehsils showed variations in trees' abundance; the maximum percentage of trees species was shown in Mandi-Bahauddin (48%), followed by Malakwal (34%) and Phalia (18%) (Figure 2A). All the values regarding the tree's distribution in Mandi-Bahauddin varied based on brick kiln distance. Mandi-Bahauddin showed variations in tree distribution, with maximum percentage (15%) noticed for *E. camaldulensis* (15%) and *V. nilotica* (15%) followed by *C. sinensis* (14%), *M. azadarach* (14%), *D. sisso* (12%) and *P. ciliate* (10%), while the lowest tree distribution was recorded in *S. cumin* (3%), *Z. manritiana* (3%) and *B. vulgaris* (2%) (Figure 2B). Phalia showed variations in tree distribution, the maximum percentage of tree species being *E. camaldulensis* (19%) and *V. nilotica* (17%) followed by *C. sinensis* (14%), *M. azadarach* (12%), *D. sisso* (10%) and *P. ciliate* (10%), while the lowest tree distribution was noticed in *A. indica* (6%), *S. cumin* (2%), *Z. manritiana* (2%) and *B. vulgaris* (2%) (Figure 2C). Malakwal showed variations in tree distribution, the maximum percentage of trees being *V. nilotica* (15%) and *E. camaldulensis* (14%), followed by *C. sinensis* (13%), *M. azadarach* (13%), *D. sisso* (10%) and *P. ciliate* (10%), while the lowest tree distribution was found in *A. indica* (6%), *S. cumin* (5%), *Z. manritiana* (3%) and *B. vulgaris* (3%) (Figure 2D).

3.2. Carbon Stock of Trees

The above ground and below ground carbon stock and total carbon stock potential of trees in Mandi-Bahauddin district were different, which describes the variations in carbon stock capacity of potential species of all three tehsils of the district (Table 4). Maximum C stock was recorded in *E. camaldulensis* (3951 Mg C ha^{-1}), followed by *S. cumin* (282 Mg C ha^{-1}), *P. ciliate* (75.9 t/ha), and minimum C stock was observed in *C. sinensis* (17.3 Mg C ha^{-1}) in Mandi-Bahauddin. In Phalia, maximum C stock was shown in *A. nilotica* (18.5 Mg C ha^{-1}), followed by *S. cumin* (16.8 Mg C ha^{-1}) and *B. vulgaris* (2.5 Mg C ha^{-1}) and minimum C stock was observed in *M. alba* (0.02 Mg C ha^{-1}). Maximum C stock was shown in *P. ciliata* (44.5 Mg C ha^{-1}), followed by *A. nilotica* (21.2 Mg C ha^{-1}), *E. camaldulensis* (20.9 Mg C ha^{-1}), and minimum C stock was observed in *M. azadarach* (0.05 Mg C ha^{-1}) in Malakwal. Maximum C stock was shown in the trees of Mandi-Bahauddin followed by Malakwal, and minimum C stock was shown in Phalia. The descending order in terms of carbon sequestration potential of trees in Mandi-Bahauddin was *S. cumin* > *E. camaldulensis* > *P. ciliate* > *B. vulgaris* > *A. nilotica* > *Z. manritiana* > *M. azadarach* > *A. Indica* > *D. sisso* > *M. alba* > *C. sinensis*. The descending order in terms of carbon sequestration potential of trees in Tehsil Phalia was *E. camaldulensis* > *A. nilotica* > *D. sisso* > *Z. manritiana* > *A. Indica* > *P. ciliate* > *S. cumin* > *M. alba* > *M. azadarach* > *C. sinensis* > *B. vulgaris*. The decreasing order in terms of carbon stock potential of trees in Malakwal was *E. camaldulensis* > *P. ciliate* > *C. sinensis* > *A. nilotica* > *Z. manritiana* > *A. Indica* > *S. cumin* > *B. vulgaris* > *D. sisso* > *M. alba* > *M. azedarach* (Table 4).

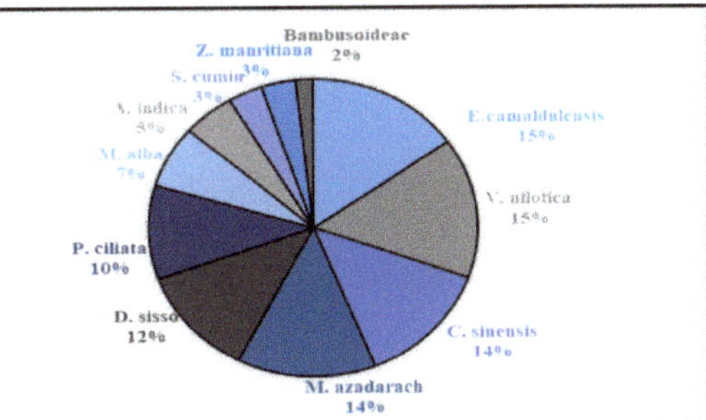

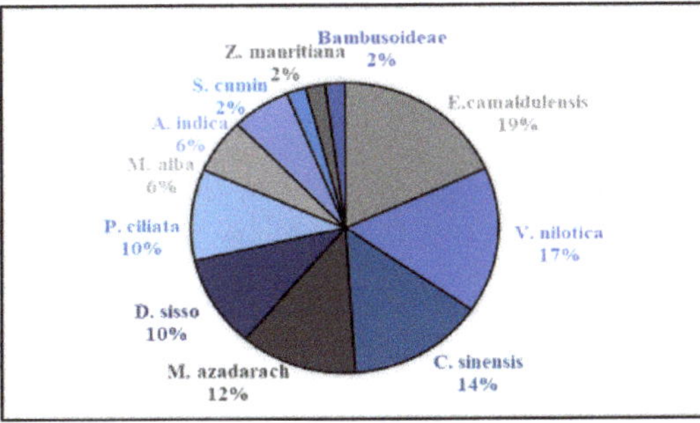

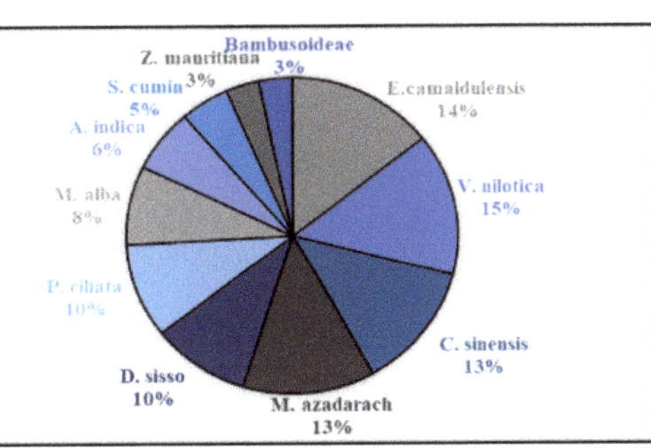

Figure 2. Tree species abundance in Mandi-Bahauddin district (**A**) and tree species distribution expressed as a percentage of total basal area for agroforestry plots in tehsils Mandi-Bahauddin (**B**), Phalia (**C**) and Malakwal (**D**) in Punjab, Pakistan.

Table 4. Summary of above ground carbon, below ground carbon and total carbon (Mg C ha^{-1}) of selected species at various points from the brick kilns.

Tehsil	Mandi-Bahauddin			Phalia			Malakwal		
Potential Tress Species	AGC	BGC	Total Carbon	AGC	BGC	Total Carbon	AGC	BGC	Total Carbon
A. nilotica	24.8	6.60	31.1	14.4	4.07	18.5	9.70	4.20	21.2
A. indica	15.0	3.90	18.9	1.30	0.60	1.29	0.06	0.02	0.12
M. azadarach	18.9	4.90	23.8	0.06	0.02	0.06	0.06	0.02	0.05
C. sinensis	13.7	3.60	17.3	0.06	0.03	0.12	0.60	0.20	0.80
D. sisso	14.9	3.90	18.9	13.3	3.50	16.8	0.07	0.03	0.09
E. camaldulensis	3135	815	3951	0.08	0.02	0.04	36.6	12.1	20.9
P. ciliata	52.0	23.6	75.9	0.20	0.09	0.40	102	60.7	44.5
S. cumin	136	35.0	282	0.50	0.13	0.98	11.5	4.40	1.20
Z. manritiana	18.0	4.70	37.8	7.10	1.65	0.13	34.0	29.8	18.7
M. alba	16.0	2.00	18.7	0.03	0.007	0.02	0.07	0.03	0.09
B. vulgaris	32.0	8.00	67.6	0.24	0.08	2.52	1.04	0.50	0.20

3.3. Total Tree Carbon Stock in Mandi-Bahauddin District

All the values regarding total tree carbon stock were different at various distances from the brick kilns in an agroforestry system in various tehsils of Mandi-Bahauddin district (Table 5). Maximum tree carbon stock (133 Mg C ha^{-1}) was noticed in Mandi-Bahauddin followed by Malakwal (62.6 Mg C ha^{-1}), while the minimum was recorded in Phalia (9.97 Mg C ha^{-1}) in (Table 5).

Table 5. Total Tree Carbon stock.

Tehsil	Total Trees C Stock (Mg C ha^{-1})
Mandi-Bahauddin	133
Phalia	9.97
Malakwal	62.6

Mean data from 3 locations of brick kilns.

3.4. Total Soil Organic Carbon Stock

Notable variations were observed for soil organic carbon at various distances from the brick kilns in the agroforestry system of various tehsils in Mandi-Bahauddin district (Table 6). Maximum total organic carbon of soil was noticed in Malakwal 10 km away from the brick kilns compared with control (near brick kilns). The highest stock of C in soil (15.30 Mg C ha^{-1}) in Mandi-Bahauddin was noticed near the brick kiln, and minimum stock of soil C (9.06 Mg C ha^{-1}) was measured 5 km away from the brick kiln, and an average organic C stock of (11.6 Mg C ha^{-1}) was noticed. In Phalia maximum measured soil C stock (13.58 Mg C ha^{-1}) was shown near the brick kiln and minimum soil C stock (11.6 Mg C ha^{-1}) was shown 5 km away from the brick kiln, and average organic carbon of (12.5 Mg C ha^{-1}) was noticed. In Malakwal, maximum soil C stock (15.3 Mg C ha^{-1}) was measured 10 km from the brick kiln and minimum soil C stock (3.12 Mg C ha^{-1}) was shown 5 km away, with an average organic C stock of (10.3 Mg C ha^{-1}) (Table 6).

Table 6. Average Soil Organic Carbon Stock.

Tehsils	Soil C Stock (Mg C ha^{-1})			Average Soil C Stock (Mg C ha^{-1})
	0 km	5 km	10 km	
Mandi-Bahauddin	15.3	9.06	10.3	11.6
Phalia	13.6	11.6	12.4	12.5
Malakwal	12.5	3.12	15.3	10.3

3.5. Total Organic Carbon Stock in Potential Crops

All the values regarding the carbon stock in Triticum aestivum and Oryza sativa at various distances were different in the agroforestry system of cultivated area for various tehsils of Mandi-Bahauddin district (Table 7). Maximum carbon stock was recorded in Triticum aestivum in Tehsil Mandi-Bahauddin (2.08×10^6 Mg C) followed by Malakwal (1.85×10^6 Mg C). Minimum carbon stock was noticed in wheat crop in Phalia (1.83×10^6 Mg C). For rice crop, maximum carbon stock was noticed in Mandi-Bahauddin (1.97×10^6 Mg C), followed by Phalia (1.83×10^6 Mg C), while minimum carbon stock in rice was recorded in Tehsil Malakwal (1.48×10^6 Mg C) (Table 8).

Table 7. Measured Carbon stock in *Triticum aestivum*.

Tehsil	AGB (Mg C ha^{-1})	BGB (Mg C ha^{-1})	Total Biomass (Mg C ha^{-1})	Total Cultivated Area (ha)	Crop C Stock (Mg C)
Mandi-Bahauddin	50.6	5.60	56.2	82,340	2.08×10^6
Phalia	52.1	5.20	57.3	71,246	1.83×10^6
Malakwal	60.8	7.10	67.8	60,754	1.85×10^6

ABG = Above ground biomass; BGB = Below ground biomass; Mean data from 3 locations of brick kilns.

Table 8. Measured Carbon stock in *Oryza sativa*.

Tehsil	AGB (Mg C ha^{-1})	BGB (Mg C ha^{-1})	Total Biomass (Mg C ha^{-1})	Total Cultivated Area (ha)	Crop C Stock (Mg C)
Mandi-Bahauddin	49.4	3.90	53.3	82,340	1.97×10^6
Phalia	50.1	4.10	57.2	71,246	1.83×10^6
Malakwal	52.3	4.90	54.2	60,754	1.48×10^6

ABG = Above ground biomass; BGB = Below ground biomass; Mean data from 3 locations of brick kilns.

4. Discussion

This study provides the estimation of potential C pools for agroforestry systems in relation to brick kilns in Mandi-Bahauddin district, Pakistan. The agroforestry systems observed in the research site are crucial for the livelihoods of local farmers, as they have both commercial and subsistence production values [36,37]. In addition to timber from tree species, field crops such as wheat also have good market prospects at local and international level, as these are widely used for consumption [38,39]. Although not intrinsically C dense compared to systems such as forests or intensively managed agroforestry systems and pastures, C storage in agricultural farms can be increased by 20.4 to 21.4 t C ha^{-1} globally [40,41] through the incorporation of long-living, deep-rooted trees [42,43]. While climatic attributes are consistent across the district where sampling was performed, the amount of C sequestered varied, largely depending on the tree species' distribution, density of tree species, basal area of tree, age of tree, area under crops and distance of agroforestry system from the brick kiln, emphasizing the importance of management decisions in determining carbon stocks. For example, district total mean tree carbon stock was the lowest (9.97 Mg ha^{-1}) in Phalia, 5 km away from the brick kiln and the highest (133 Mg ha^{-1}) in Mandi-Bahauddin, 10 km away from the brick kiln, among all tehsils in the study area. This appears to be related primarily to the level of tree stocking in the district, with Mandi-Bahauddin having the highest average basal area of all tehsils (8.9 m^2 ha^{-1}), and Malakwal having the lowest tree basal area (3.2 m^2 ha^{-1}). The current study revealed that maximum carbon sequestration was noticed in *E. camaldulensis* while the minimum was observed in *M. alba*. Among various tree species, the difference in biomass might be due to numerous aspects, i.e., number of trees/ha, tree age, quality and location of site, cultural practices, techniques and system of planting and ecological conditions in that area [21–24]. The average stored C stock of agroforestry systems in our study area was 11.46 Mg C ha^{-1} in soil, 68.5 Mg C ha^{-1} in trees, 60.4 Mg C ha^{-1} in wheat and 54.2 Mg C ha^{-1} in rice. The carbon stocks of simple systems (i.e., combination of single trees with cash crop or grass) and mixed-tree systems are similar to C stock on agroforestry systems reported by other studies in Indonesia [44,45]. Tree biomass accumulation representing the value of tree basal area has a correlation with C stock value [46]. The tree species having a higher basal area have the capacity for higher biomass accumulation which results in higher C stocks [47]. Our findings reflect that the plots near the brick kilns showed more growth of trees and potential crops (wheat and rice) in the study area, due to shifting of brick kilns to zig-zag technology that reduced dust emission. Similar patterns were reported in other studies [21,36,37,48,49].

In a terrestrial ecosystem, soil is a very notable system for CO_2 mitigation. Many agroecologists have revealed that the soil carbon pool has prime importance in agroforestry systems [50,51]. The outcomes of the current investigation regarding SOC supports the hypothesis that SOC contents are maximum at 0–15 cm layers, with increased higher buildup of tree litter [52]. Higher soil carbon contents in agroforestry systems largely depend on the quality and amount of biomass input by non-tree and tree components of the system [53,54]. Moreover, vegetation detritus and litter from pruning under proper agroforestry management return a greater amount of organic carbon to the soil [55,56] (Figure 3). In this study, the average SC pools ranged from 10.3 to 15.3 Mg C ha^{-1} 10 km away from the brick kilns. Kimaro et al. [57] reported that a significant variation was noticed in carbon sequestered by an agroforestry system with legume trees, compared with a mono-crop system. Carbon storage in plants can be high in complex agroforestry systems and productivity of field crops depends on several factors such as age, structure and the way the systems are managed [58,59]. The results are comparable with the findings of other studies [42,60–62], which reported that agroforestry can store carbon in the range of 12–228 Mg ha^{-1}. In our study it was noticed that maximum tree C stock potential (133 Mg C ha^{-1}) was observed in Mandi-Bahauddin followed by Malakwal (62.6 Mg C ha^{-1}), 10 km away from the brick kilns, while the minimum value was observed in Phalia (9.97 Mg C ha^{-1}), 5 km away from brick kilns. In our research, the carbon sequestration potential of an agroforestry system (soil, trees and crops) showed differential response in relation to brick kilns (5 and 10 km away from brick kilns). Similar observations were reported by Gera et al. [63] who claimed that the variations in the carbon sequestration potential relate to the mean annual increment, which varies with site, age, density and plantation, as well as with the quality of planting stock.

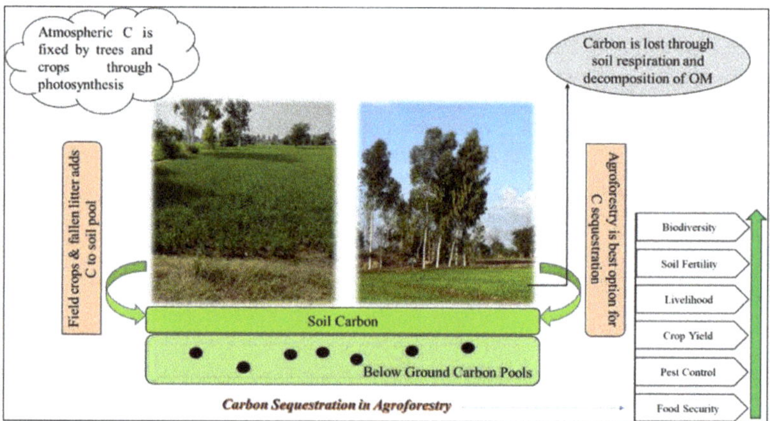

Figure 3. Schemes of carbon sequestration in agroforestry.

Various factors such as species, land use type, cultural practices and CO_2 supply play important roles in C stock and C sequestration rate [64]. Most farmers planted trees on their farmland for a short rotation. After regular intervals, harvesting of cultivated trees results in loss of C, but C is again stored when the harvested wood is converted into plywood, packaging materials, poles and the manufacturing of furniture [65]. Moreover, more accumulation of biomass was observed at the early stages and this decreased with the passage of time and age [66]. Tree stem sequesters the carbon for a longer time after felling as compared to the carbon stored in leaves and branch biomass [59]. These results highlight both the current and potential carbon sequestration potential of agroforestry in Pakistan, and can be further used in devising strategies for implementing tree planting programs on agricultural land and designing future carbon sequestration projects in Pakistan.

5. Conclusions

For the development of better management strategies, understanding of the influence of potential trees on farmlands in linear plantations is critical. Our intensive sampling in three tehsils showed that agroforestry systems in Punjab, Pakistan, currently store moderate amounts of carbon in plants and soil. In the agroforestry system, the increased soil organic carbon was due to litter fall and gave higher monetary returns in terms of more C stock 5 and 10 km away from the brick kilns. The decreasing order in terms of C stock potential of crops was wheat > rice. Variation was observed for C stock in crops (wheat and rice) at various distances from the brick kiln at the sampling sites. More C stock in the crops was noticed in Mandi-Bahauddin due to the maximum cultivated area, followed by Phalia and Malakwal. Crops had the maximum potential to store carbon dioxide at different sampling points, compared to trees and soil. The descending order in terms of C stock potential of agroforestry was crops > trees > soil. The findings of this study suggest that planting of tree species along with farm crops is a sustainable way to mitigate climate change by sequestering large amounts of carbon from the atmosphere. However, future studies should be conducted to highlight more indicators associated with the operation of the brick kilns. Given appropriate incentives, Punjab's farmers could help Pakistan meet its commitments to the Paris Climate Accord through reasonable changes in tree planting on existing agroforestry systems.

Author Contributions: Conceptualization, Q.u.Z., G.Y. and Y.C.; methodology, N.K., I.T. and S.N.; software, G.Y.; formal analysis, N.K.; investigation, Q.u.Z.; resources, N.K.; data curation, K.A.; writing—original draft preparation, N.K., M.A., M.W. and A.B.; writing—review and editing, Q.u.Z. and Y.C.; supervision, Q.u.Z.; project administration, Q.u.Z. and I.T. All authors have read and agreed to the published version of the manuscript.

Funding: This research received no external funding.

Institutional Review Board Statement: Not applicable.

Informed Consent Statement: Not applicable.

Data Availability Statement: Not Applicable.

Acknowledgments: We are thankful to Umair Riaz (Scientific Officer), Soil and Water Testing Laboratory for Research Bahawalpur, Punjab Government, Pakistan for providing the research facilities for soil analysis.

Conflicts of Interest: The authors declare no conflict of interest.

References

1. Eckstein, D.; Künzel, V.; Schäfer, L. Global Climate Risk Index 2021. Who Suffers Most from Extreme Weather Events 2021, 2000–2019. Available online: https://www.germanwatch.org/en/19777 (accessed on 25 January 2021).
2. Mondal, M.; Goswami, S.; Ghosh, A.; Oinam, G.; Tiwari, O.N.; Das, P.; Halder, G.N. Production of biodiesel from microalgae through biological carbon capture: A review. *3 Biotech* **2017**, *7*, 99. [CrossRef]
3. Khan, M.A.; Khan, J.A.; Ali, Z.; Ahmad, I.; Ahmad, M.N. The challenge of climate change and policy response in Pakistan. *Environ. Earth Sci.* **2016**, *75*, 412. [CrossRef]
4. Davenport, F.V.; Diffenbaugh, N.S. Using machine learning to analyze physical causes of climate change: A case study of US Midwest extreme precipitation. *Geophys. Res. Lett.* **2021**, *48*, e2021GL093787. [CrossRef]
5. Dalby, S. Global climate change and security threats. In *Handbook of Security and the Environment*; Edward Elgar Publishing: Washington, DC, USA, 2021.
6. World Bank. *World Development Report 2008: Agriculture for Development*; Herndon, V.A., Ed.; World Bank Publisher: Washington, DC, USA, 2017; pp. 20172–20960.
7. Reilly, J.; Tubiello, F.; McCarl, B.; Abler, D.; Darwin, R.; Fuglie, K.; Rosenzweig, C. US agriculture and climate change: New results. *Clim. Chang.* **2003**, *57*, 43–67. [CrossRef]
8. Van Ginkel, M.; Biradar, C. Drought Early Warning in Agri-Food Systems. *Climate* **2021**, *9*, 134. [CrossRef]
9. Jahan, S.; Falah, S.; Ullah, H.; Ullah, A.; Rauf, N. Antioxidant enzymes status and reproductive health of adult male workers exposed to brick kiln pollutants in Pakistan. *Environ. Sci. Pollut. Res.* **2016**, *23*, 12932–12940. [CrossRef]
10. Monga, V.; Singh, L.P.; Bhardwaj, A.; Singh, H. Respiratory health in brick kiln workers. *Int. J. Phys. Soc. Sci.* **2012**, *2*, 226–244.
11. Raza, A.; Qamer, M.F.; Afsheen, S.; Adnan, M.; Naeem, S.; Atiq, M. Particulate matter associated lung function decline in brick kiln workers of Jalalpur Jattan, Pakistan. *Pak. J. Zool.* **2014**, *46*, 237–243.
12. Islam, M.S.; Akbar, A.; Akhtar, A.; Kibria, M.M.; Bhuyan, M.S. Water quality assessment along with pollution sources of the Halda River. *J. Asiat. Soc. Bangladesh Sci.* **2017**, *43*, 61–70. [CrossRef]
13. Ullah, S.; Ahmad, W.; Majeed, M.T.; Sohail, S. Asymmetric effects of premature deagriculturalization on economic growth and CO2 emissions: Fresh evidence from Pakistan. *Environ. Sci. Pollut. Res.* **2021**, *28*, 66772–66786. [CrossRef]
14. Vignola, R.; Harvey, C.A.; Bautista-Solis, P.; Avelino, J.; Rapidel, B.; Donatti, C.; Martinez, R. Ecosystem-based adaptation for smallholder farmers: Definitions, opportunities and constraints. *Agric. Ecosyst. Environ.* **2015**, *211*, 126–132. [CrossRef]
15. Abbas, F.; Hammad, H.M.; Fahad, S.; Cerdà, A.; Rizwan, M.; Farhad, W.; Bakhat, H.F. Agroforestry: A sustainable environmental practice for carbon sequestration under the climate change scenarios a review. *Environ. Sci. Pollut. Res.* **2017**, *24*, 11177–11191. [CrossRef]

16. Arévalo-Gardini, E.; Farfán, A.; Barraza, F.; Arévalo-Hernández, C.O.; Zúñiga-Cernades, L.B.; Alegre, J.; Baligar, V.C.C. Growth, Physiological, Nutrient-Uptake- Efficiency and Shade-Tolerance Responses of Cacao Genotypes under Different Shades. *Agronomy* **2021**, *11*, 1536. [CrossRef]
17. Hidayat, N.; Sianipar, J. The Potential of Agroforestry in Supporting Food Security for Peatland Community—A Case Study in the Kalampangan Village, Central Kalimantan. *Ecol. Eng.* **2021**, *22*, 123–130.
18. Murthy, I.K.; Gupta, M.; Tomar, S.; Munsi, M.; Tiwari, R.; Hegde, G.T.; Ravindranath, N.H. Carbon sequestration potential of agroforestry systems in India. *J. Earth Sci. Clim. Chang.* **2013**, *4*, 1–7. [CrossRef]
19. Saco, P.M.; McDonough, K.R.; Rodriguez, J.F.; Rivera-Zayas, J.; Sandi, S.G. The role of soils in the regulation of hazards and extreme events. *Philos. Trans. R. Soc. Lond. B* **2021**, *376*, 20200178. [CrossRef]
20. Baig, M.B.; Burgess, P.J.; Fike, J.H. Agroforestry for healthy ecosystems: Constraints, improvement strategies and extension in Pakistan. *Agrofor. Syst.* **2021**, *95*, 995–1013. [CrossRef]
21. Yasin, G.; Nawaz, M.F.; Martin, T.A.; Niazi, N.K.; Gul, S.; Yousaf, M.T.B. Evaluation of Agroforestry Carbon Storage Status and Potential in Irrigated Plains of Pakistan. *Forests* **2019**, *10*, 640. [CrossRef]
22. Giri, K.; Pandey, R.; Jayaraj, R.S.C.; Nainamalai, R.; Ashutosh, S. Regression equations for estimating tree volume and biomass of important timber species in Meghalaya, India. *Curr. Sci.* **2019**, *116*, 75–81. [CrossRef]
23. Ravindranath, N.H.; Ostwald, M. Methods for Below-Ground Biomass. In *Carbon Inventory Methods: Handbook for Greenhouse Gas Inventory, Carbon Mitigation and Roundwood Production Projects*; Springer Science & Business Media: Dordrecht, The Netherlands, 2008; Volume 29, pp. 149–156.
24. Thomas, S.C.; Martin, A.R. Carbon content of tree tissues. A Synthesis. *Forests* **2012**, *3*, 332–352. [CrossRef]
25. Rawat, V.R.S.; Kishwan, J. Forest conservationbased, climate changemitigation approach for India. *Int. For. Rev.* **2008**, *10*, 269–280. [CrossRef]
26. Brown, S.; Gillespie, A.J.; Lugo, A.E. Biomass estimation methods for tropical forests with applications to forest inventory data. *For. Sci.* **1989**, *35*, 881–902.
27. Cairns, M.A.; Brown, S.; Helmer, E.H.; Baumgardner, G.A. Root biomass allocation in the world's upland forests. *Oecologia* **1997**, *111*, 1–11. [CrossRef]
28. Hawkins, T. *Biomass and Volume Tables for Eucalyptus camaldulensis, Dalbergia sissoo, Acacia auriculiformis and Cassia siamea in the Central Bhabar-Terai of Nepal*; Oxford Forestry Institute, University of Oxford: Oxford, UK, 1987; p. 33.
29. Roy, M.M.; Pathak, P.S.; Rai, A.K.; Kushwaha, D. Tree growth and biomass production in Melia azedarach on farm boundaries in a semi-arid region. *Indian For.* **2006**, *132*, 105–110.
30. Das, D.K.; Chaturvedi, O.P. Structure and function of Populus deltoides agroforestry systems in eastern India: 1. Dry matter dynamics. *Agrofor. Syst.* **2005**, *65*, 215–221. [CrossRef]
31. Rai, S.N. Above ground biomass in tropical rain forests of Western Ghats, India. *Indian For.* **1984**, *110*, 754–764.
32. Singh, B.; Tripathi, K.P.; Singh, K. Community structure, diversity, biomass and net production in a rehabilitated subtropical forest in north India. *Open J. For.* **2011**, *1*, 11. [CrossRef]
33. Walkley, A.J.; Black, I.A. An examination of the Degtjare method for determining soil organic matter and a proposed modification of the chromic acid titration method. *Soil Sci.* **1934**, *37*, 29–38. [CrossRef]
34. De Joa Carlos, M.S.; Carlos, C.C.; Warren, A.D.; Lal, R.; Filho, S.P.V.; Piccolo, M.C.; Feigl, B.E. Organic matter dynamics and carbon sequestration rates for a tillage chrono sequence in a Brazilian Oxisol. *Soil Sci. Soc. Am. J.* **2001**, *65*, 1486–1499.
35. Prommer, J.; Walker, T.W.; Wanek, W.; Braun, J.; Zezula, D.; Hu, Y.; Richter, A. Increased microbial growth, biomass, and turnover drive soil organic carbon accumulation at higher plant diversity. *Glob. Chang. Biol. Bioenergy* **2020**, *26*, 669–681. [CrossRef]
36. Siarudin, M.; Rahman, S.A.; Artati, Y.; Indrajaya, Y.; Narulita, S.; Ardha, M.J.; Larjavaara, M. Carbon Sequestration Potential of Agroforestry Systems in Degraded Landscapes in West Java, Indonesia. *Forests* **2021**, *12*, 714. [CrossRef]
37. Khadka, D.; Aryal, A.; Bhatta, K.P.; Dhakal, B.P.; Baral, H. Agroforestry Systems and Their Contribution to Supplying Forest Products to Communities in the Chure Range, Central Nepal. *Forests* **2021**, *12*, 358. [CrossRef]
38. Razafindratsima, O.H.; Kamoto, J.F.; Sills, E.O.; Mutta, D.N.; Song, C.; Kabwe, G.; Sunderland, T. Reviewing the evidence on the roles of forests and tree-based systems in poverty dynamics. *Policy Econ.* **2021**, *131*, 102576. [CrossRef]
39. Ranjan, R. Payments for ecosystems services-based agroforestry and groundwater nitrate remediation: The case of Poplar deltoides in Uttar Pradesh, India. *J. Clean. Prod.* **2021**, *287*, 125059. [CrossRef]
40. Zomer, R.J.; Neufeldt, H.; Xu, J.; Ahrends, A.; Bossio, D.; Trabucco, A.; Wang, M. Global Tree Cover and Biomass Carbon on Agricultural Land: The contribution of agroforestry to global and national carbon budgets. *Sci. Rep.* **2016**, *6*, 29987. [CrossRef] [PubMed]
41. Grigorov, B.; Assenov, A. Tree Cover and Biomass Carbon on Agricultural Land in Mala Planina. In *Smart Geography*; Springer: Cham, Switzerland, 2020; pp. 265–274.
42. Albrecht, A.; Kandji, S.T. Carbon sequestration in tropical agroforestry systems. *Agric. Ecosyst. Environ.* **2003**, *99*, 15–27. [CrossRef]
43. Schwarz, J.; Schnabel, F.; Bauhus, J. A conceptual framework and experimental design for analysing the relationship between biodiversity and ecosystem functioning (BEF) in agroforestry systems. *Basic Appl. Ecol.* **2021**, *56*, 51–60. [CrossRef]

44. Rahayu, S.; Lusiana, B.; van Noordwijk, M. Above ground carbon stock assessment for various land use systems in Nunukan, East Kalimantan. In *Carbon Stock Monitoring in Nunukan, East Kalimantan: A Spatial and Modelling Approach. Report from the Carbon Monitoring Team of the Forest Resource Management for Carbon Sequestration (FORMACS) Project*; Lusiana, B., van Noordwijk, M., Rahayu, S., Eds.; World Agroforestry Center: Bogor, Indonesia, 2005; pp. 21–34.
45. Zhu, X.; Liu, W.; Chen, J.; Bruijnzeel, L.A.; Mao, Z.; Yang, X.; Jiang, X.J. Reductions in water, soil and nutrient losses and pesticide pollution in agroforestry practices: A review of evidence and processes. *Plant Soil* **2020**, *453*, 45–86. [CrossRef]
46. Sajad, S.; Haq, S.M.; Yaqoob, U.; Calixto, E.S.; Hassan, M. Tree composition and standing biomass in forests of the northern part of Kashmir Himalaya. *Vegetos* **2021**, *34*, 1–10. [CrossRef]
47. Laskar, S.Y.; Sileshi, G.W.; Nath, A.J.; Das, A.K. Allometric models for above and below-ground biomass of wild Musa stands in tropical semi evergreen forests. *Glob. Ecol. Conserv.* **2020**, *24*, e01208. [CrossRef]
48. Marone, D.; Poirier, V.; Coyea, M.; Olivier, A.; Munson, A.D. Carbon storage in agroforestry systems in the semi-arid zone of Niayes, Senegal. *Agrofor. Syst.* **2017**, *91*, 941–954. [CrossRef]
49. Sanogo, K.; Dayamba, D.S.; Villamor, G.B.; Bayala, J. Impacts of Climate Change on Ecosystem Services of Agroforestry Systems in the West African Sahel: A Review. *Agrofor. Deg. Landsc.* **2020**, *4*, 213–224.
50. Huang, L.; Liu, J.; Shao, Q.; Xu, X. Carbon sequestration by forestation across China: Past, present and future. *Renew. Sust. Energy Rev.* **2012**, *16*, 1291–1299. [CrossRef]
51. Nawaz, M.F.; Mazhar, K.; Gul, S.; Ahmad, I.; Yasin, G.; Asif, M.; Tanvir, M. Comparing the early stage carbon sequestration rates and effects on soil physicochemical properties after two years of planting agroforestry trees. *J. Basic Appl. Sci.* **2017**, *13*, 527–533.
52. Kaushal, R.; Verma, K.S.; Chaturvedi, O.P.; Alam, N.M. Leaf litter decomposition and nutrient dynamics in four important multiple tree species. *Range Manag. Agrofor.* **2012**, *33*, 20–27.
53. Sarto, M.V.; Borges, W.L.; Sarto, J.R.; Rice, C.W.; Rosolem, C.A. Deep soil carbon stock, origin, and root interaction in a tropical integrated crop–livestock system. *Agrofor. Syst.* **2020**, *94*, 1865–1877. [CrossRef]
54. Hairiah, K.; van Noordwijk, M.; Sari, R.R.; Saputra, D.D.; Suprayogo, D.; Kurniawan, S.; Gusli, S. Soil carbon stocks in Indonesian (agro) forest transitions: Compaction conceals lower carbon concentrations in standard accounting. *Agric. Ecosyst. Environ.* **2020**, *294*, 106879. [CrossRef]
55. Ma Howald, N.M.; Randerson, J.T.; Lindsay, K.; Munoz, E.; Doney, S.C.; Lawrence, P.; Hoffman, F.M. Interactions between land use change and carbon cycle feedbacks. *Glob. Biogeochem. Cycles* **2017**, *31*, 96–113. [CrossRef]
56. Hübner, R.; Kühnel, A.; Lu, J.; Dettmann, H.; Wang, W.; Wiesmeier, M. Soil carbon sequestration by agroforestry systems in China: A meta-analysis. *Agric. Ecosyst. Environ.* **2021**, *315*, 107437. [CrossRef]
57. Kimaro, A.A.; Isaac, M.E.; Chamsharma, S.A.O. Carbon pools in tree biomass and soils under rotational woodlot systems in eastern Tanzania. In *Carbon Sequestration Potential of Agroforestry Systems: Opportunities and Challenges*; Kumar, B.M., Nair, P.K.R., Eds.; Springer Science + Business Media: Dordrecht, The Netherlands, 2010; Volume 8, pp. 129–144.
58. Oelbermann, M.; Voroney, R.P.; Gordon, A.M. Carbon sequestration in tropical and temperate agroforestry systems: A review with examples from Costa Rica and Southern Canada. *Agric. Ecosyst. Environ.* **2004**, *104*, 359–377. [CrossRef]
59. Swamy, S.L.; Puri, S. Biomass production and C-sequestration of Gmelina arborea in plantation and agroforestry system in India. *Agrofor. Syst.* **2005**, *64*, 181–195. [CrossRef]
60. Rizvi, R.H.; Dhyani, S.K.; Yadav, R.S.; Singh, R. Biomass production and carbon stock of popular agroforestry systems in Yamunanagar and Saharanpur districts of northwestern India. *Curr. Sci.* **2011**, *100*, 736–742.
61. Wani, N.; Velmurugan, A.; Dadhwal, V.K. Assessment of agricultural crop and soil carbon pools in Madhya Pradesh, India. *Trop. Ecol.* **2010**, *51*, 11–19.
62. Yadav, R.P.; Bisht, J.K. Agroforestry: A way to conserve MPTs in North Western Himalaya. *Res. J. Agric. For. Sci.* **2013**, *1*, 8–13.
63. Gera, M.; Mohan, G.; Bisht, N.S.; Gera, N. Carbon sequestration potential of agroforestry under CDM in Punjab state of India. *Indian J. For.* **2011**, *34*, 1–10.
64. Ramachandran Nair, P.K.; Mohan Kumar, B.; Nair, V.D. Agroforestry as a strategy for carbon sequestration. *J. Plant Nutr. Soil Sci.* **2009**, *172*, 10–23. [CrossRef]
65. Arora, G.; Chaturvedi, S.; Kaushal, R.; Nain, A.; Tewari, S.; Alam, N.M.; Chaturvedi, O.P. Growth, biomass, carbon stocks, and sequestration in an age series of Populus deltoides plantations in Tarai region of central Himalaya. *Turk. J. Agric. For.* **2014**, *38*, 550–560. [CrossRef]
66. Yasin, G.; Nawaz, M.F.; Siddiqui, M.T.; Niazi, N.K. Biomass, carbon stocks and CO_2 sequestration in three different aged irrigated *Populus deltoides* bartr. Ex marsh. Bund planting agroforestry systems. *Appl. Ecol. Environ. Res.* **2018**, *16*, 6239–6252. [CrossRef]

 *agriculture*

Article

Liming and Phosphate Application Influence Soil Carbon and Nitrogen Mineralization Differently in Response to Temperature Regimes in Allophanic Andosols

Chihiro Matsuoka-Uno [1], Toru Uno [1], Ryosuke Tajima [1], Toyoaki Ito [1,2] and Masanori Saito [1,3,*]

[1] Field Science Center, Graduate School of Agricultural Science, Tohoku University, Osaki 989-6711, Japan; c.matsuoka91@gmail.com (C.M.-U.); uno@tohoku.ac.jp (T.U.); tazy@tohoku.ac.jp (R.T.); toyoaki-ito@nafu.ac.jp (T.I.)

[2] Food Industry Department, Niigata Agro-Food University, Niigata 959-2702, Japan

[3] Faculty of Agriculture, Iwate University, Morioka 020-8550, Japan

* Correspondence: masanori.saito.b6@tohoku.ac.jp

Abstract: Andosols are characterized by high organic matter content and play a significant role in carbon storage. However, they have low phosphorus fertility because of the high phosphate-fixing capacity of active aluminum. For agricultural use of Andosols, it is necessary to ameliorate its poor phosphorus fertility by applying lime and high doses of phosphate fertilizers. The objective of the present study was to clarify how such soil amendments affect the mineralization of soil organic carbon (C) and nitrogen (N) in allophanic Andosols under different temperature regimes. The soil was treated using combinations of liming and heavy phosphate application, followed by incubation under different temperature conditions. The N mineralization and the soil CO_2 evolution rate were measured periodically. The patterns of N mineralization were analyzed by fitting them to first-order kinetics. Liming increased C and N mineralization irrespective of temperature, and the increase was further enhanced by phosphate application. Kinetic analysis of the N mineralization curve indicated lowering of the activation energy of N mineralization reactions with phosphate application, suggesting that P application may accelerate N mineralization at lower temperatures. These findings provide a basis for developing soil management strategies to reduce the loss of soil organic matter.

Keywords: Andosols; soil carbon; global warming; phosphorus; phosphate; liming; nitrogen; mineralization; temperature

Citation: Matsuoka-Uno, C.; Uno, T.; Tajima, R.; Ito, T.; Saito, M. Liming and Phosphate Application Influence Soil Carbon and Nitrogen Mineralization Differently in Response to Temperature Regimes in Allophanic Andosols. *Agriculture* **2022**, *12*, 142. https://doi.org/ 10.3390/agriculture12020142

Academic Editor: Laura Zavattaro

Received: 9 December 2021
Accepted: 20 January 2022
Published: 21 January 2022

Publisher's Note: MDPI stays neutral with regard to jurisdictional claims in published maps and institutional affiliations.

1. Introduction

A huge amount of carbon (C), 1500 Pg C, approximately twice the amount present in the atmosphere and thrice the amount in terrestrial plant biomass, is stored in the soil as organic matter [1]. Furthermore, any variation in the soil organic matter content significantly impacts the global carbon cycle. In arable soils, the amount of soil C is strongly influenced by environmental factors such as temperature and agricultural management practices such as tillage. Life-cycle inventory analysis of greenhouse gases (GHGs) in various crop production systems in northern Japan revealed that carbon dioxide (CO_2) emissions from soil were the largest source of GHG emissions [2]. Various reports have warned that the increase in decomposition of soil organic matter due to global warming will further accelerate the atmospheric CO_2 concentration [3–5]. In order to take counter-measures against global warming, it is essential to predict the potential changes in soil C status in response to environmental changes and human activities (such as agricultural soil management practices).

Andosols are of volcanic origin and are one of the most widely distributed soils in Japan (31% of the country's land area and 47% of upland arable fields) [6]. Andosols are characterized by high organic matter content, resulting in high soil C storage capacity and excellent physical and water retention properties. However, they have low phosphorus fertility because of the high phosphate-fixing capacity of active aluminum. For agricultural use of Andosols in Japan, when uncultivated Andosols are converted into arable lands, it is recommended to ameliorate the soil by applying a mixture of phosphate fertilizers (e.g., fused phosphate and superphosphate) equivalent to 10% of the phosphate absorption coefficient (PAC) of the soil [7]. Such large amounts of phosphate fertilization not only increase the amount of available phosphate in the soil, but also increase phosphorus fertility of the soil by masking some of the phosphate fixing sites of clay particles and decreasing the phosphorus adsorption capacity [6,8]. These amendments may increase the activity of soil microorganisms that were previously limited by the phosphorous deficiency and acidity of the soil and may promote the decomposition of soil organic matter. Munevar and Wollum [9] reported that the addition of phosphate and N increased the mineralization of soil organic C and N in the Andosols of Colombia. Ogasawara et al. [10] reported that soil N mineralization increased with the addition of a mixture of fused phosphate and superphosphate in the Andosols of northern Japan. These incubation experiments were conducted at about 30 °C. However, in the northern part of Japan under cool-temperate climate, the average annual soil temperature is 10–15 °C and rarely exceeds 25 °C, even in summer. Although a large number of studies on temperature dependency of soil organic matter decomposition have been reported, e.g., [3–5], the effect of soil amendments on soil organic matter decomposition focusing on temperature regime has rarely been reported.

The objective of the present study was to clarify how these soil amendments affect mineralization of soil organic C and N, with an emphasis on temperature response. We hypothesize that soil available phosphate limits C and N mineralization in an allophanic Andosol and that this effect is more evident in lower temperature regime. In the present study, we used a soil sample from an uncultivated land with an allophanic Andosols under cool-temperate climate (annual mean temperature 10.6 °C) and treated the soil with different combinations of lime and phosphate. Phosphate was applied at the rate of 5% and 20% of PAC, as indicated in the study by Yamamoto and Miyasato [7]. Mineralization of soil organic carbon and nitrogen was examined by incubating the soil under different temperature regimes.

2. Materials and Methods

2.1. Soil

The soil was collected from the surface layer (0–20 cm) of an uncultivated land (39°44′56.4″ N, 141°08′16.8″ E) under *Pinus densiflora* trees in Morioka City, Iwate Prefecture, Japan, after removing the litter layer of soil. The physicochemical properties of the soil were as follows: texture, LiC; total C, 186 g C kg^{-1} dry soil; total N, 11.7 g N kg^{-1} dry soil; ammonium N, 63 mg N kg^{-1} dry soil; nitrate N, 65 mg N kg^{-1} dry soil; pH (H$_2$O), 5.3; PAC (phosphate absorption coefficient), 23.3 g P$_2$O$_5$ kg^{-1} dry soil; available phosphate, 0.016 g P$_2$O$_5$ kg^{-1} dry soil (Truog method); maximum water holding capacity, 0.80 cm^3/cm^3. Analytical methods of soil are described in the Supplementary Materials. The soil was classified as an allophanic cumulic Andosol according to the Soil Classification System of Japan [6].

2.2. Design for Incubation Experiment

Fresh, moist soil samples that were air-dried, sieved through a 2 mm mesh, and treated with different levels of phosphate and lime were used for the incubation experiment. In the non-liming treatment, two levels of potassium phosphate monobasic (0% for control, 5% and 20% of PAC; equivalent to about 11.6 Mg P$_2$O$_5$ ha^{-1}, 46.6 MgP$_2$O$_5$ ha^{-1}) were added to the soil samples. In the liming treatment, calcium hydroxide (5.5 g kg^{-1} dry soil) was used to adjust the soil pH (to ~6.0) before adding potassium phosphate at the rate of 0%, 5%, and 20% of PAC. Soil sample was incubated for each treatment in triplicate (see the Supplementary Materials). The incubation temperatures were 20 and 30 °C. An additional temperature level, 25 °C, was set for N mineralization in the non-liming treatment, because preliminary experiment showed another temperature level may be required for more reliable kinetic parameters of N mineralization (Section 2.4) in the non-liming treatment.

The treated soil samples were put in separate 100 mL polyethylene vials, individually covered with aluminum foil, and incubated for 120 days. Throughout the incubation period, the soil moisture condition was maintained at 50% of the maximum water holding capacity of the soil. The pH (H_2O) and available phosphate content (Truog method) of the soil samples were measured periodically throughout the incubation period.

2.3. C Mineralization: CO_2 Evolution

Organic C mineralization was estimated by periodically measuring the rate of CO_2 evolution from the soil (Figure S1). The aluminum foil was removed from the incubated samples, and the polyethylene vials were individually placed in 485 mL wide-mouthed glass bottles with a septum for gas sampling. The bottles were incubated at the respective incubation temperatures. About 1 h and 24 h after the bottles were incubated, 1 mL gas was collected from the bottle using a gas-tight syringe, and the CO_2 concentration was measured using a gas chromatograph (GC-8A, Shimadzu, Kyoto, Japan) equipped with a thermal conductivity detector and a Shincarbon ST column (Shinwa Chemical Industries Ltd., Kyoto, Japan). The CO_2 evolution rate was calculated from the increase in the CO_2 concentration during this period. The soils in the polyethylene vials after CO_2 assay were back to the incubator for further CO_2 evolution measurement.

2.4. N Mineralization

After the designated incubation period, mineral N was extracted from the soil samples using 2 N KCl. The nitrate-N, nitrite-N, and ammonium-N contents of each extract were estimated using an autoanalyzer (QuAAtro 2-HR, BL-TECH Co., Tokyo, Japan), and the sum of these N values was considered as the mineral N content of the respective sample.

2.5. Kinetic Analysis of N Mineralization Curve

The mineral N contents obtained in this study were presented in the form of a N mineralization curve, which was analyzed using the method proposed by Sugihara et al. [11]. This method assumes that the N mineralization curve corresponds to an equation of the first-order reaction (Equation (1)), and the relationship between the mineralization rate constant and temperature follows the Arrhenius law (Equation (2)).

$$N_t = N_0\{1 - \exp(-k \cdot t)\} \tag{1}$$

$$k = A \cdot \exp(-Ea/R \cdot T) \tag{2}$$

where N_t and N_0, amount of mineralized N at time t and infinite time, respectively (mg N kg^{-1} soil); k, mineralization rate constant (day^{-1}); Ea, activation energy (kJ mol^{-1}); T, absolute temperature (K); R, universal gas constant (8.31 J mol^{-1} K^{-1}); A, Arrhenius constant.

Curve fitting of the data was performed using the Solver program of Microsoft Excel 2019 (Microsoft Corporation, Redmond, WA, USA). Fitness was evaluated on the basis of the minimum Akaike information criterion values. Parameters obtained by this non-liner curve fitting were compared among the treatments by Bonfferoni method (see the Supplementary Materials).

2.6. Statistical Analyses

Statistical analyses were conducted using the JMP software (SAS Institute Japan, Tokyo, Japan) and the data analyses tools of Microsoft Excel 2019.

3. Results

3.1. Change of Available Phosphate and pH of Soil during Incubation

The changes in pH are shown in Figure 1. The pH of soil samples subjected to liming treatment was initially adjusted to ≈6.0. In all treatments, the pH gradually decreased, probably due to the accumulation of nitrate N, while a small increase in pH during the early incubation period was found in some treatments. Overall, the change in pH during the incubation period was within one unit.

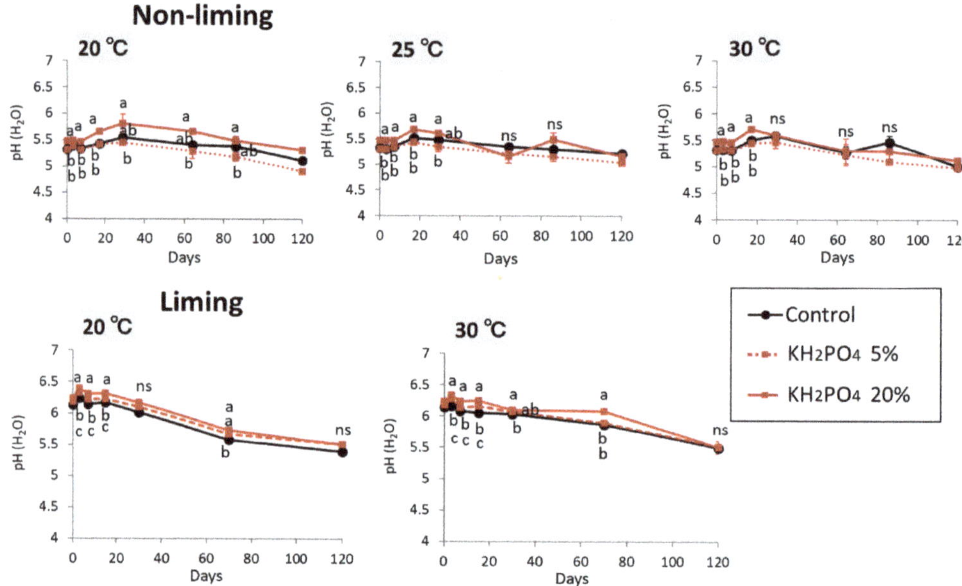

Figure 1. Change of soil pH (H_2O) during the incubation period. Vertical bars indicate the standard error of means (n = 3). Different letters indicate significantly different on the same day (Tukey's multiple comparison test, ns; not significantly different). The data on day 120 in non-liming treatment were not replicated, and therefore the statistical analyses were not performed.

The change in the available phosphate at 20 °C is shown in Figure 2. The trend at 30 °C was almost the same as that observed at 20 °C (data not shown). Available phosphate decreased immediately after the addition of potassium phosphate due to the quick fixation of phosphate, and then remained constant, for both application levels. In the liming treatment, the amount of available phosphate was higher than that in the non-liming treatment. It may have been due to a somewhat higher pH in the liming treatment.

3.2. CO_2 Evolution

The change in the rate of CO_2 evolution from the soil during the incubation period is shown in Figure 3. The pattern of changes in the CO_2 evolution rate differed between the non-liming and liming treatments. In the non-liming treatment, the CO_2 evolution rate increased with time, peaked at 40–50 days, and then decreased. However, liming treatment increased the overall CO_2 evolution rate of the soil, compared to the non-liming treatment. There was a peak in CO_2 evolution at the beginning of the incubation period, followed by a decline and a small peak at ≈40 days, similar to that in the non-liming treatment. The total CO_2 evolution during the incubation period (Figure 4) was estimated by integrating the curves of the CO_2 evolution rate up to 84 days. In the non-liming treatment, phosphate application was not significant in terms of the total amount of CO_2 evolved. In contrast, in the liming treatment, phosphate application increased total CO_2 evolution at 30 °C (Figure 4).

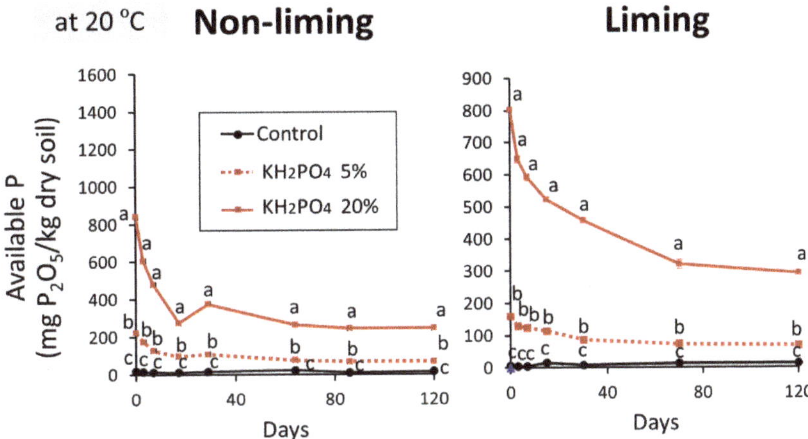

Figure 2. Change of available phosphate in soil (Truog method) at 20 °C. Vertical bars indicate the standard error of means (*n* = 3). Different letters indicate significantly different on the same day (Tukey's multiple comparison test).

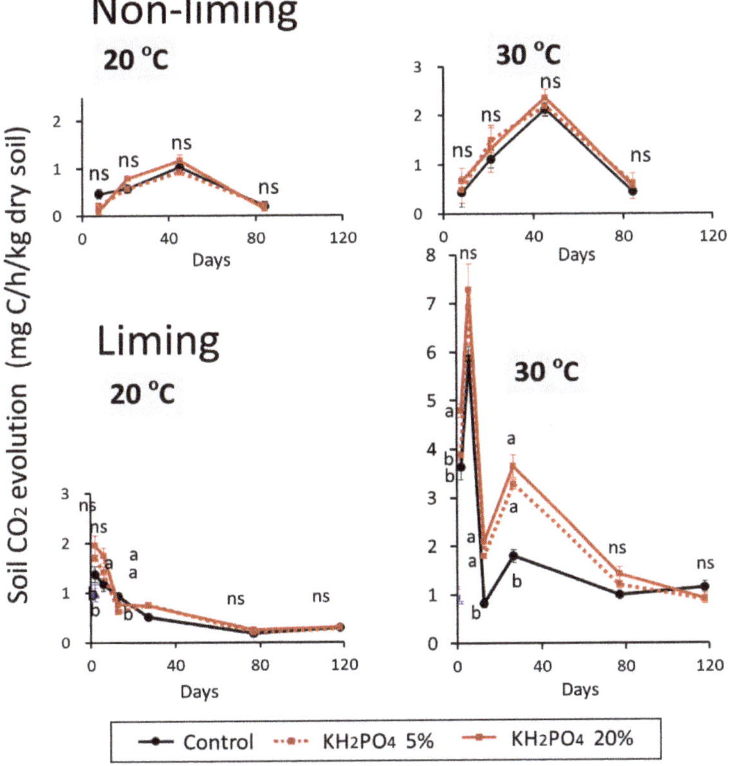

Figure 3. CO$_2$ evolution rates at different temperatures. Vertical bars indicate the standard error of means (*n* = 3). Different letters indicate significantly different at each temperature (Tukey's multiple comparison test; ns, not significantly different).

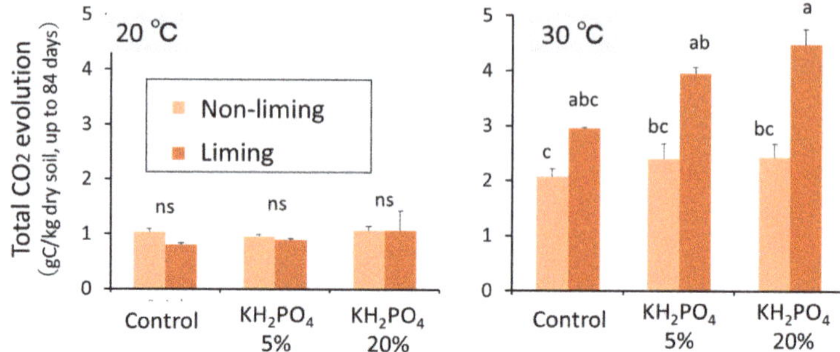

Figure 4. Total C mineralization estimated up to 84 days by CO_2 evolution rate at different temperatures. Vertical bars indicate the standard error of means (n = 3). Different letters indicate significantly different at each temperature (Tukey's multiple comparison test; ns, not significantly different). Two ways ANOVA (temperature × liming) showed both temperature and liming treatments showed $p < 0.01$. Since the CO_2 evolution data at 20 °C were obtained only up to day 84 (Figure 3), total CO_2 evolution at 30 °C was also estimated up to day 84. The data for day 84 in the liming treatment was extrapolated from the data for day 77 and day 120.

3.3. N Mineralization Curves and Their Kinetic Analysis

The accumulation of mineral N in the soil during the incubation period is shown in Figure 5. In the non-liming treatment, N mineralization at 20 °C increased with the application of phosphate at the rate of 20% PAC. In the liming treatment, N mineralization increased to some degree for both levels of phosphate application. The N mineralization parameters were obtained by fitting the N mineralization curve to the first-order reaction (Table 1). The mineralization rate constants were not significantly affected by both phosphate application and liming. The activation energy (Ea) in the non-liming treatment significantly decreased with phosphate application (20% of PAC). Liming increased the amount of potentially mineralizable N (N_0).

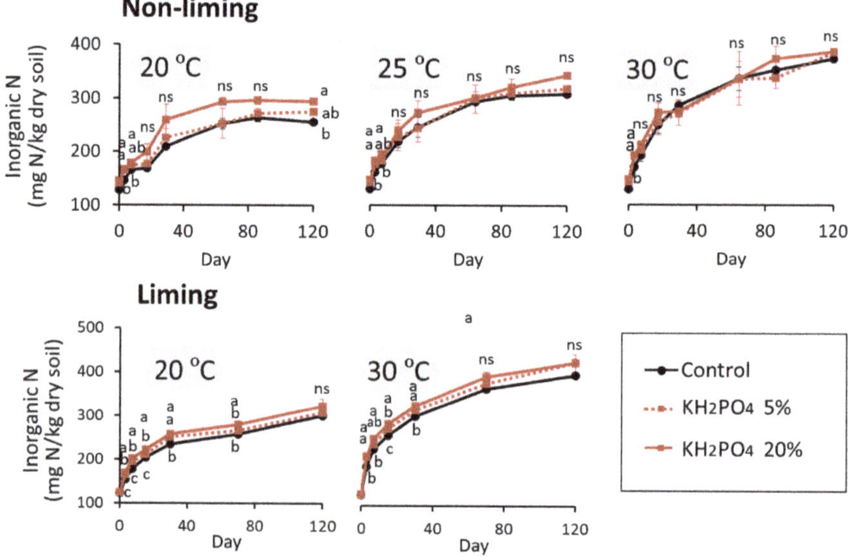

Figure 5. Change of mineral N in soils at different incubation treatments and temperatures. Vertical bars indicate the standard error of means (n = 3). Different letters indicate significant difference at each temperature (Tukey's multiple comparison test; ns, not significantly different).

Table 1. Nitrogen mineralization parameters of soil incubated at different treatments and temperatures.

Treatment		No *		k *		Ea *	
Liming	Phosphate	mg N/kg Dry Soil		/day, 25 °C		kJ/mol	
		#		#			#
	Control	200		0.022		5.39	a
Non-liming	KH_2PO_4 5%	178	ns	0.021	ns	5.03	a
	KH_2PO_4 20%	188		0.029		3.45	b
	(average)	(189)		(0.024)		(4.62)	
	Control	228		0.018		5.92	
Liming	KH_2PO_4 5%	240	ns	0.016	ns	6.13	ns
	KH_2PO_4 20%	237		0.019		6.06	
	(average)	(235)		(0.018)		(6.04)	
	##	*		ns		ns	

* No, potentially mineralizable N; k, mineralization rate constant; Ea, activation energy. # Different letters (a, b) indicate statistical significance among phosphate addition treatments by the Bonferroni method ($p < 0.05$, ns: not significant). ## Statistical difference between the non-liming and liming treatments (*: $p < 0.05$, ns: not significant).

4. Discussion

In the present study, we examined the effects of liming and heavy phosphate application on C and N (N) mineralization in soil samples from the uncultivated allophanic Andosols of Japan.

The effect of liming showed the same trend irrespective of the temperature. Liming increased the rate of CO_2 evolution during the early stages of incubation (Figure 3). The N mineralization curve also showed the same trend—liming increased N mineralization in the initial stages of incubation. This result was supported by the kinetic analysis of the N mineralization curve, which showed that N mineralization potential, N_0, was increased by liming (Table 1).

In contrast, the effect of phosphate application varied with liming and incubation temperature. The total CO_2 evolution, which is the integrated value of the CO_2 evolution during the incubation period (Figure 4), increased with phosphate application and was more pronounced at 30 °C in the liming treatments, whereas it was not as evident in the non-liming treatments. N mineralization in the non-liming treatment at 20 °C was promoted by phosphate application. N mineralization parameters based on the kinetic analysis of N mineralization curves showed that the activation energy, Ea, in the non-liming treatment decreased with an increase in the phosphate application rate (Table 1).

The enhancement of soil organic matter decomposition by liming is well documented [12–16]. This reaction occurs partly because of the increase in microbial activity due to the neutralization of soil acidity, and partly because of the solubilization of organic matter. Liming releases organic matter trapped in soil aggregates by the dispersion of clay. Takahashi et al. [17] reported that the addition of lime to Andosols broke down a part of the Al–humus complex and led highly humified soil organic matter to more decomposable. Another source of solubilized organic matter might be microbial biomass. Marumoto et al. [18] reported that lime application killed some microbial biomass (especially fungi) and caused a rapid growth of microorganisms that utilize the dead microbial biomass. The microorganisms responsible for mineralization immediately after liming—as observed in the early stages of incubation in the present study—are probably the newly proliferated microorganisms, "fast growers" that utilize the solubilized and easily degradable fraction of organic matter. These "fast growers" may be more active at a higher temperature (30 °C), and more phosphate may be required for their activity (Figure 4).

The relationship between the activation energy (Ea) of N mineralization and the average available soil phosphate during the incubation period (Figure 6) shows that the increase in available phosphate leads to a decrease in Ea in the non-liming treatment, indicating that the mineralization reaction occurs more readily at lower temperatures. In terms of temperature dependency of organic matter decomposition, recalcitrant organic matter may show higher activation energy and thus may show its higher temperature dependency than labile organic matter [19]. In contrast to liming as discussed above, it is unlikely that the addition of phosphate may have had a significant impact on organic matter quality. In the non-liming treatments, the microorganisms involved in C and N mineralization may be "slow growers" that decompose indigenous soil organic matter. Their activity is probably slower than that of the "fast growers" observed during the early stages of liming treatments. The microorganisms might have adapted to temperatures lower than 30 °C, and their activity in this temperature range might have been enhanced by phosphate addition.

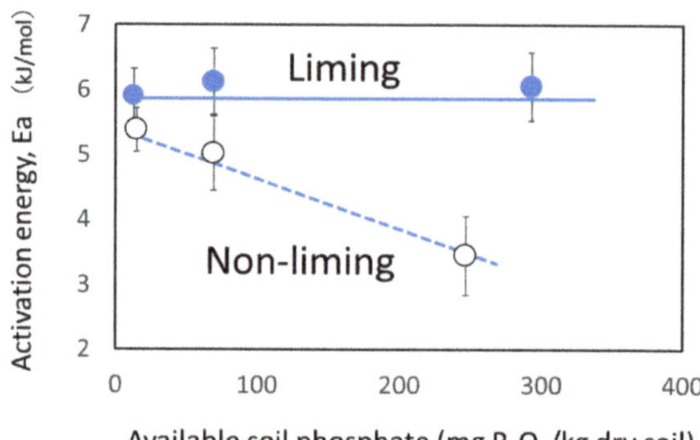

Figure 6. Relationship between apparent activation energy of nitrogen mineralization and available soil phosphate. The available soil phosphate (Troug method) indicated here is the average of the available soil phosphate values during the incubation period (15–120 days, 20 °C). Bars indicate the standard error of means ($n = 3$). Lines in figure were arbitrarily drawn.

Several studies have reported that soil C and N mineralization is limited by the phosphate availability in highly weathered tropical soils with high phosphate-fixing capacity [20,21] and in cool, temperate soils rich in organic content, such as Andosols [9,22–24]. These studies did not refer to temperature dependency of the mineralization. Saito [25] examined the N mineralization characteristics of various soils in the northern Tohoku region of Japan and found a negative correlation between available p and Ea. This corresponds with the present result in non-liming treatment (Figure 6). Microbial populations at lower temperatures than the optimum require higher concentrations of substrates due to lowering affinity to substrate [26]. In the control soil in the non-liming treatment, available phosphate was kept very low (Figure 2), and therefore that phosphate may be deficit for microorganisms. The addition of phosphate may mitigate such phosphate deficit for soil microorganism and may result in lowering Ea. Microbial community analysis of a wide range of Japanese Andosols showed that fungal diversity decreased with increasing soil available phosphate, and *Mortiellra* tended to be one of the dominant fungal groups [27]. This also might be related to changes in the temperature responsiveness of N mineralization to phosphate application, as observed in Ea (Figure 6).

Global warming is threatening to accelerate the loss of soil organic carbon. In fact, the thawing of frozen soils in polar regions is becoming a reality [28,29]. This may be also a serious issue in cool and cold temperate regions, where soils are rich in organic matter. In Japan, higher doses of phosphate fertilizers are used in agricultural production, partly because Andosols, which have a high phosphate-fixing capacity, are widely distributed in the arable lands. This practice has increased the available phosphate in arable soils [30]. Intensive fertilization of arable soils may accelerate the loss of soil organic carbon.

For sustainable agricultural production in the changing climatic conditions, soil management practices must be improved to facilitate soil carbon storage [31]. Many studies have been conducted for this purpose [32,33]. The present study indicates that the effect of intensive fertilizer management on organic matter decomposition should be investigated more systematically and extensively, as well as being reflected in sustainable soil management.

5. Conclusions

Conversion of uncultivated Andosols to arable lands generally requires liming treatments and heavy phosphate applications to improve crop production [6,8]. The effects of liming and heavy application of phosphate on soil C and N mineralization in allophanic Andosols were investigated under different temperature regimes. Liming led highly humified organic matter to more decomposable [16,17,24] and increased the easily degradable fraction of organic matter, and as a result, the microbial communities that proliferated using these fractions became more active with the application of high amounts of phosphate. In contrast, in the non-liming treatment, phosphate increased soil N mineralization at lower temperatures. These findings support our hypothesis raised in the Introduction section and also provide a basis for the development of a model for predicting soil C and N mineralization. The model may contribute to the development of management strategies to reduce the loss of soil organic matter in the face of global warming.

Supplementary Materials: The following are available online at https://www.mdpi.com/article/10.3390/agriculture12020142/s1, Analytical methods of soil, Incubation of soil, Curve fitting of N mineralization curve. Figure S1: Measurement of soil CO_2 evolution rate.

Author Contributions: Conceptualization, M.S.; methodology, C.M.-U., T.U., R.T., T.I. and M.S.; data analysis, C.M.-U., R.T. and M.S.; writing, C.M.-U. and M.S. All authors have read and agreed to the published version of the manuscript.

Funding: This research was in-part supported by ESPEC Foundation for Global Environment Research and Technology (ESPEC Prize for the Encouragement of Environmental Studies).

Data Availability Statement: The data presented in this study are available on request from the corresponding author.

Acknowledgments: We are grateful to Shintaro Hara for his technical support of gas chromatographic analysis.

Conflicts of Interest: The authors declare no conflict of interest.

References

1. IPCC. *Climate Change 2001: The Scientific Basis*; Cambridge University Press: Cambridge, UK, 2001; pp. 1–944.
2. Koga, N.; Sawamoto, T.; Tsuruta, H. Life cycle inventory-based analysis of greenhouse gas emissions from arable land farming systems in Hokkaido, northern Japan. *Soil Sci. Plant Nutr.* **2006**, *52*, 564–574. [CrossRef]
3. Crowther, T.W.; Todd-Brown, K.E.; Rowe, C.W.; Wieder, W.R.; Carey, J.C.; Machmuller, M.B.; Snoek, B.L.; Fang, S.; Zhou, G.; Allison, S.D.; et al. Quantifying global soil carbon losses in response to warming. *Nature* **2016**, *540*, 104–108. [CrossRef]
4. Davidson, E.A.; Janssens, I.A. Temperature sensitivity of soil carbon decomposition and feedbacks to climate change. *Nature* **2006**, *440*, 165–173. [CrossRef]
5. Von Lützow, M.; Kögel-Knabner, I. Temperature sensitivity of soil organic matter decomposition—What do we know? *Biol. Fertil. Soils* **2009**, *46*, 1–15. [CrossRef]
6. Takata, Y. 4.3 Andosols. In *The Soils of Japan*; Hatano, R., Shinjo, H., Takata, Y., Eds.; Springer: Singapore, 2021; pp. 83–90. [CrossRef]
7. Yamamoto, T.; Miyasato, S. Studies on the soil productivity of upland fields. Heavy application of phosphate mixture in Iwate volcanic ash soil. *Bull. Tohoku Agric. Exp. Stn.* **1971**, *42*, 53–92.
8. Shoji, S.; Nanzyo, M.; Dahlgren, R. Chapter 8 Productivity and Utilization of Volcanic Ash Soils. In *Volcanic Ash Soils Genesis, Properties and Utilization, Developments in Soil Science, Volume 21*; Shoji, S., Nanzyo, M., Dahlgren, R., Eds.; Elsevier: Amsterdam, The Netherlands, 1993; pp. 209–251. [CrossRef]
9. Munevar, F.; Wollum, A.G. Effects of the addition of phosphorus and inorganic nitrogen on carbon and nitrogen mineralization in Andepts from Colombia. *Soil Sci. Soc. Am. J.* **1977**, *41*, 540–545. [CrossRef]
10. Ogaswara, K.; Yamamoto, T. Studies on dynamics of nitrogen in upland soils. (Part 8) On application of nitrogen fertilizer with heavy application of phosphate. *Tohoku Agric. Res.* **1967**, *8*, 146–149.

11. Sugihara, S.; Konno, T.; Ishii, K. Kinetics of mineralization of organic nitrogen in soil. *Bull. Natl. Inst. Agro-Environ. Sci.* **1986**, *1*, 127–166.

12. Chan, K.Y.; Heenan, D.P. Lime-induced loss of soil organic carbon and effect on aggregate stability. *Soil Sci. Soc. Am. J.* **1999**, *63*, 1841–1844. [CrossRef]

13. Curtin, D.; Campbell, C.A.; Jalil, A. Effects of acidity on mineralization: pH-dependence of organic matter mineralization in weakly acidic soils. *Soil Biol. Biochem.* **1997**, *30*, 57–64. [CrossRef]

14. Haynes, R.J.; Naidu, R. Influence of lime, fertilizer and manure applications on soil organic matter content and soil physical conditions: A review. *Nutr. Cycl. Agroecosyst.* **1998**, *51*, 123–137. [CrossRef]

15. Li, Y.; Wang, T.; Camps-Arbestain, M.; Suárez-Abelenda, M.; Whitby, C.P. Lime and/or phosphate application affects the stability of soil organic carbon: Evidence from changes in quantity and chemistry of the soil water-extractable organic matter. *Environ. Sci. Tech.* **2020**, *54*, 13908–13916. [CrossRef]

16. Paradelo, R.; Virto, I.; Chenu, C. Net effect of liming on soil organic carbon stocks: A review. *Agric. Ecosyst. Environ.* **2015**, *202*, 98–107. [CrossRef]

17. Takahashi, T.; Ikeda, Y.; Fujita, K.; Nanzyo, M. Effect of liming on organically complexed aluminum of nonallophanic Andosols from northeastern Japan. *Geoderma* **2006**, *130*, 26–34. [CrossRef]

18. Marumoto, T.; Okano, S.; Nishio, T. Effect of liming on the mineralization of microbial biomass nitrogen in volcanic grassland soil. *Soil Microorg.* **1990**, *36*, 5–10.

19. Wagai, R.; Kishimoto-Mo, A.; Yonemura, S.; Shirato, Y.; Hiradate, S.; Yagasaki, Y. Linking temperature sensitivity of soil organic matter decomposition to its molecular structure, accessibility, and microbial physiology. *Glob. Chang. Biol.* **2013**, *19*, 1114–1125. [CrossRef]

20. Cleveland, C.C.; Townsend, A.R.; Schmidt, S.K. Phosphorus limitation of microbial processes in moist tropical forests: Evidence from short-term laboratory incubations and field studies. *Ecosystems* **2002**, *5*, 680–691. [CrossRef]

21. Cleveland, C.C.; Townsend, A.R. Nutrient additions to a tropical rain forest drive substantial soil carbon dioxide losses to the atmosphere. *Proc. Nat. Acad. Sci. USA* **2006**, *103*, 10316–10321. [CrossRef] [PubMed]

22. Kranabetter, J.M.; Banner, A.; de Groot, A. An assessment of phosphorus limitations to soil nitrogen availability across forest ecosystems of north coastal British Columbia. *Can. J. For. Res.* **2005**, *35*, 530–540. [CrossRef]

23. Kevan, J.M.; Melany, C.F.; Peter, M.G. Calcium and phosphorus interact to reduce mid-growing season net nitrogen mineralization potential in organic horizons in a northern hardwood forest. *Soil Biol. Biochem.* **2011**, *43*, 271–279.

24. Miyazawa, M.; Takahashi, T.; Sato, T.; Kanno, H.; Nanzyo, M. Factors controlling accumulation and decomposition of organic carbon in humus horizons of Andosols. *Biol. Fertil. Soils* **2013**, *49*, 929–938. [CrossRef]

25. Saito, M. Nitrogen mineralization parameters and its availability indices of soils in Tohoku district, Japan: Their relationship. *Jpn. J. Soil Sci. Plant Nutr.* **1990**, *61*, 265–272.

26. Nedwell, D.B. Effect of low temperature on microbial growth: Lowered affinity for substrates limits growth at low temperature. *FEMS Microb. Ecol.* **1999**, *30*, 101–111. [CrossRef]

27. Bao, Z.; Matsushita, Y.; Morimoto, S.; Hoshino, Y.T.; Suzuki, C.; Nagaoka, K.; Takenaka, M.; Murakami, H.; Kuroyanagi, Y.; Urashima, Y.; et al. Decrease in fungal biodiversity along an available phosphorous gradient in arable Andosol soils in Japan. *Can. J. Microbiol.* **2013**, *59*, 368–373. [CrossRef]

28. IPCC. Climate Change 2021. The Physical Science Basis. Summary for Policymakers. 2021. Available online: https://www.ipcc.ch/report/ar6/wg1/downloads/report/IPCC_AR6_WGI_SPM_final.pdf (accessed on 1 December 2021).

29. Natali, S.M.; Holdren, J.P.; Rogers, B.M.; Treharne, R.; Duffy, P.B.; Pomerance, R.; MacDonald, E. Permafrost carbon feedbacks threaten global climate goals. *Proc. Nat. Acad. Sci. USA* **2021**, *118*, e2100163118. [CrossRef]

30. Obara, H.; Nakai, M. Available phosphate of arable lands in Japan. Changes of soil characteristics in Japanese Arable Lands (II). *Jpn. J. Soil Sci. Plant Nutr.* **2004**, *75*, 59–67. [CrossRef]

31. Rattan, L. Managing soils for negative feedback to climate change and positive impact on food and nutritional security. *Soil Sci. Plant Nutr.* **2020**, *66*, 1–9. [CrossRef]

32. Bell, S.M.; Barriocanal, C.; Terrer, C.; Rosell-Melé, A. Management opportunities for soil carbon sequestration following agricultural land abandonment. *Environ. Sci. Policy* **2020**, *108*, 104–111. [CrossRef]

33. Bai, X.; Huang, Y.; Ren, W.; Coyne, M.; Jacinthe, P.A.; Tao, B.; Hui, D.; Yang, J.; Matocha, C. Responses of soil carbon sequestration to climate-smart agriculture practices: A meta-analysis. *Glob. Chang. Biol.* **2019**, *25*, 2591–2606. [CrossRef] [PubMed]

 *agriculture*

MDPI

Article

Source and Accumulation of Soil Carbon along Catena Toposequences over 12,000 Years in Three Semi-Natural *Miscanthus sinensis* Grasslands in Japan

David S. Howlett [1], J. Ryan Stewart [2], Jun Inoue [3], Masanori Saito [4], DoKyoung Lee [5], Hong Wang [6], Toshihiko Yamada [7], Aya Nishiwaki [8], Fabián G. Fernández [9] and Yo Toma [10,*]

[1] United States Agency for International, 1300 Pennsylvania Avenue, NW, Washington, DC 20523, USA; davhowlett@gmail.com

[2] Department of Plant and Wildlife Sciences, Brigham Young University, 2124 LSB, Provo, UT 84602, USA; rstewart@byu.edu

[3] Graduate School of Science, Osaka City University, 3-138 Sugimoto, Sumiyoshi-ku, Osaka 558-8585, Japan; juni@sci.osaka-cu.ac.jp

[4] Graduate School of Agricultural Science, Tohoku University, Osaki 989-6711, Japan; masanori.saito.b6@tohoku.ac.jp

[5] Department of Crop Sciences, University of Illinois at Urbana-Champaign, Turner Hall, 1102 S Goodwin Ave, Urbana, IL 61801, USA; leedk@illinois.edu

[6] Interdisciplinary Research Center of Earth Science Frontier, Beijing Normal University, Beijing 100875, China; hongwang@bnu.edu.cn

[7] Field Science Center for Northern Biosphere, Hokkaido University, Kita-11, Nishi-10, Kita-ku, Sapporo 060-0811, Japan; yamada@fsc.hokudai.ac.jp

[8] Field Science Center Kibana Agricultural Science Station, Faculty of Agriculture, University of Miyazaki, 1-1, Gakuen-Kibanadai-Nishi, Miyazaki 889-2192, Japan; nishiwaki@cc.miyazaki-u.ac.jp

[9] Department of Soil, Water and Climate, University of Minnesota, 1529 Gortner Avenue, St. Paul, MN 55108, USA; fabiangf@umn.edu

[10] Research Faculty of Agriculture, Hokkaido University, Kita-9, Nishi-9, Kita-ku, Sapporo 060-8589, Japan

* Correspondence: toma@chem.agr.hokudai.ac.jp; Tel.: +81-11-706-3857

Citation: Howlett, D.S.; Stewart, J.R.; Inoue, J.; Saito, M.; Lee, D.; Wang, H.; Yamada, T.; Nishiwaki, A.; Fernández, F.G.; Toma, Y. Source and Accumulation of Soil Carbon along Catena Toposequences over 12,000 Years in Three Semi-Natural *Miscanthus sinensis* Grasslands in Japan. *Agriculture* **2022**, *12*, 88. https://doi.org/10.3390/agriculture12010088

Academic Editor: Claudia Di Bene

Received: 21 November 2021
Accepted: 31 December 2021
Published: 10 January 2022

Publisher's Note: MDPI stays neutral with regard to jurisdictional claims in published maps and institutional affiliations.

Abstract: *Miscanthus*-dominated semi-natural grasslands in Japan appear to store considerable amounts of soil C. To estimate the long-term effect of *Miscanthus* vegetation on the accumulation of soil carbon by soil biota degradation in its native range, we measured total soil C from the surface to a 1.2 m depth along a catena toposequence in three annually burned grasslands in Japan: Kawatabi, Soni, and Aso. Soil C stock was estimated using a radiocarbon age and depth model, resulting in a net soil C accumulation rate in the soil. C_4-plant contribution to soil C accumulation was further estimated by $\delta^{13}C$ of soil C. The range of total soil C varied among the sites (i.e., Kawatabi: 379–638 Mg, Soni: 249–484, and Aso: 372–408 Mg C ha^{-1}). Catena position was a significant factor at Kawatabi and Soni, where the toe slope soil C accumulation exceeded that of the summit. The soil C accumulation rate of the whole horizon in the grasslands, derived C mainly from C_4 plant species, was 0.05 ± 0.02 (Average $\pm$ SE), 0.04 ± 0.00, and 0.24 ± 0.04 Mg C ha^{-1} yr^{-1} in Kawatabi, Soni, and Aso, respectively. Potential exists for long-term sequestration of C under *M. sinensis*, but the difference in the C accumulation rate can be influenced by the catena position and the amount of vegetation.

Keywords: *Miscanthus sinensis*; soil carbon; catena; radiocarbon dating; C_4 grasses

1. Introduction

Miscanthus, a cold-tolerant perennial grass C_4 native to East and Southeast Asia, exhibits potential as a feedstock for the production of biofuels and bio-based products [1–4]. As such, this genus may see a considerable increase in cultivation in the United States and Europe in the coming years. In order to estimate the potential effects on edaphic resources, several researchers have considered the impact this genus has on soil carbon, mostly in cultivated or fallow fields [5–9]. Observations, however, from these studies have been limited to less than 20 years. Semi-natural *Miscanthus sinensis* grasslands in Japan, some of which have been managed for hundreds of years [4], offer an opportunity to assess the effects of centuries of *Miscanthus* growth and management on soil C resources [10–15].

Miscanthus utilizes the efficient C_4 photosynthetic pathway and, consequently, the origin of organic inputs to the soil from plants in this genus can be determined via their stable isotopic composition [9,16,17]. Differences in the relative abundance of the $^{13}C/^{12}C$ ratio in plants that utilize the C_4 photosynthetic pathway allow for determining the relative contribution by *Miscanthus* to soil C stocks [18]. Using stable C isotopic analysis, Schneckenberger and Kuzyakov (2007) estimated *Miscanthus* C inputs ranging between 0.11 and 0.30 g C kg soil^{-1} yr^{-1} in a sandy versus loamy soil in Germany. Howlett et al. (2013) found that a majority of soil C, ranging between 52% and 85% at a depth up to 1.5 m, was derived from *Miscanthus* in a Typic Melanudans in a southern Japanese *Miscanthus*-dominated grassland. To estimate soil C accumulation over time, the relationship between soil depth and age needs to be determined. Dating of soil C in soil profiles with low C content, however, is problematic. Bioturbation causes vertical mixing or movement of soluble C compounds, and additions of heterogeneous sources of C from land surface and groundwater reduce the integrity of age-to-depth models [19,20]. However, soils previously investigated in the same biome contained high C content and demonstrated highly correlated age-to-depth models ($R^2 = 0.98$–0.99) [11]. Using these age-to-depth models, Howlett et al. (2013) estimated *Miscanthus*-derived soil C accumulation at 0.62–0.85 Mg C ha^{-1} yr^{-1} down to a 1.5 m depth in a *Miscanthus*-dominated semi-natural grassland in southern Japan.

In volcanic regions with diverse topography where ash accumulation and other formative materials, such as humus, are subject to erosion and deposition processes along the continuum of different landscape positions [21], a soil catena study that encompasses hillside summits, mid-slopes, and toe slopes can shed valuable information on C accumulation phenomena. The dynamics of C associated with these topographic forms have previously been considered. Schimel et al. (1985) [22] reported double the surface soil C on lower foot slopes relative to that found in summit soils in Colorado, USA. While fully vegetated grasslands may not experience a considerable amount of erosion, the annual burning associated with traditionally managed *Miscanthus* grasslands in Japan removes the vegetative cover that protects soil from erosion [11,13]. Precipitation events following these traditional burnings redistributes C-containing sediments lower into the watershed, and soils developed on lower positions on a slope may demonstrate higher levels of soil C due to the fact of depositional processes. As soil from organic horizons erode and C-containing sediment accumulates below, the catena concept provides utility in characterizing the potential variability of soil C within the varied topography of many Japanese *Miscanthus* grasslands. Understanding the variability and accumulation of soil C underlying *Miscanthus* grasslands in Japan will help to determine whether there is greater benefit in growing *Miscanthus* as a soil C sequestration bioenergy crop or to serve alternative purposes such as a traditional landscape for ecotourism. Our efforts to identify the long-term impacts of *Miscanthus* on soil C deposition involved a three-pronged approach: (1) estimate the effect of catena position on the development of C stocks to 1.2 m underlying three semi-natural grassland catenas currently dominated by *M. sinensis* in Japan, (2) quantify the relative contribution of *M. sinensis* to these C stocks, and (3) calculate the rate of soil C accumulation contributed by *M. sinensis*.

2. Materials and Methods

2.1. Site Descriptions

Three grasslands sites were chosen based on their current dominance by *M. sinensis* and for a latitudinal gradient across Japan (Figure 1a).

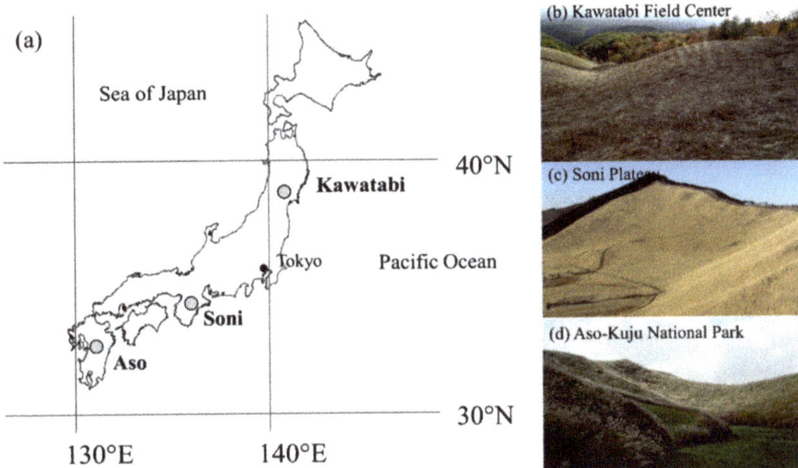

Figure 1. Locations of the semi-natural *Miscanthus sinensis* grassland study sites in Japan (**a**) and the semi-natural *Miscanthus* grassland study sites of the Kawatabi Field Science Center, Miyagi Prefecture (**b**); Soni Plateau, Nara Prefecture (**c**); Aso-Kuju National Park, Kumamoto Prefecture (**d**).

2.1.1. Kawatabi

The northernmost sampling site was at the Kawatabi Field Science Center of Tohoku University, located near the Kawatabi natural springs (KAW), Miyagi Prefecture, Japan (38°46.25′ N, 140°45.16′ E, 550 m a.s.l., approximately 14° of the slope), where the mean annual temperature is 11 °C, and there is a mean annual precipitation of 1460 mm [23] (Figure 1b). The site is dominated by *M. sinensis*, which is maintained by annual mechanical cutting in the fall with the grass left in place after cutting. Burning as a maintenance practice in this site ceased more than 40 years ago. Ito and Saigusa (1996) described several soil profiles from this site. One key aspect of this site is the documented non-allophanic chemistry of the Andisols. High contents of Al and Fe help to retain relatively large amounts of organic matter facilitating the formation of stable organo–mineral complexes, which likely also occur at the other two sites.

2.1.2. Soni Plateau

The middle latitude site was within the grasslands at the Soni Plateau (SONI) in Nara Prefecture, Japan (34°31.07′ N, 136°09.80′ E, 720 m a.s.l. approximately 30° of the slope), where the mean annual temperature is 12 °C, and it has a mean annual precipitation of 1720 mm (Figure 1c). The *M. sinensis*-dominated grasslands is maintained by annual burning in the spring, and the area is a tourist destination for recreation and ecotourism. Soils are also Typic Andisols with contents of volcanic glass [24]. Inoue et al. (2012) [24] and Okunaka et al. (2012) [25] reconstructed the vegetative history of the site, with *M. sinensis* becoming dominant on the site 1500 years ago.

2.1.3. Aso-Kuju

The southern-most site is within the grasslands of the Aso-Kuju National Park (ASO) in Kumamoto Prefecture, Japan (32°55.75′ N, 131°09.60′ E, 843 m a.s.l., approximately 21° of the slope), where the mean annual temperature is 13 °C, and it has a mean annual precipitation of 3200 mm (Figure 1d). The *M. sinensis*-dominated site is considered a semi-natural grassland, which has been annually burned for hundreds of years in early spring in order to maintain the *Miscanthus*-grassland ecosystem [4,13,26]. No additional management has taken place other than burning for at least 50 years [11]. The *Miscanthus* grasslands of

Aso-Kuju National Park are a tourist destination partly due to the rarity of grasslands in Japan. Moreover, ecotourism during the burning season significantly contributes to the local economy. Soils in this region are typical of Japan, derived from volcanic ash and characterized by the USDA soil classification system as Typic Melanudans [27]. A diagnostic general characteristic of these soils is the presence of a 2AB (K-Ah) volcanic deposit at approximately 60–70 cm depths in many parts of the caldera, which has been dated to a local volcanic eruption event by Mount Kikai approximately 7300 years ago [15,28].

2.2. Field Collections

At each of the three sites, soils were sampled at 10 cm increments down to 120 cm, following a hillside catena sequence transect, representing the summit, mid-, and toe slope. Transect lengths were 200–250 m from the toe slope to the summit. Replicate soil samples were taken 3 m on the left- and right-hand sides of each catena location (facing the summit). We selected the east and west slope aspects at each site to avoid known edaphic differences between north- and south-facing slopes. The slopes of the catena sequences varied among the sites, and the mean percent slope was calculated by dividing vertical distance from the summit to the toe slope by the transect distance. The mean percent slopes were 29% in KAW, 36% in SONI, and 18% in ASO. In total, 27 soil catenas were examined (three sites, three catena positions, and three replicates per catena position).

To estimate bulk density, a metal canister of a known volume (i.e., 100 cm^3) was inserted into the soil profile and removed with an undisturbed soil core at four representative depths (15, 45, 75, and 105 cm, which were the midpoints for 0–30, 30–60, 60–90, and 90–120 cm soil depths). Bulk density samples were dried at 100 °C to a consistent weight.

2.3. Laboratory Procedures

Soil samples from each of the 10 cm depth increments for soil C content analysis were immediately stored in temperature-controlled conditions and then air-dried to a constant weight. Air-dried soils were passed through a 2 mm sieve, and subsamples were taken for determination of moisture content. Soil C content was determined by combustion of 2 mm sieved soil in an elemental analyzer (Leco CN Analyzer, St. Joseph, MI, USA).

Stable isotopic C composition and accelerator mass spectrometer (AMS) radiocarbon ages were determined for five soil depth increments: 0–10, 20–30, 50–60, 80–90, and 110–120 cm at each of the catena positions at all three sites. Soil samples with high organic C content were pretreated using the standard acid–base–acid (ABA) method as described by Brandt et al. (2012) [29]. The same pretreatment method was also applied to radiocarbon-free wood, IAEA (International Atomic Energy Agency) C5 wood, and FIRI-D (Fifth International Radiocarbon Inter-Comparison D) woodworking standards. Approximately 0.5 g of soil, 3–5 mg of working standards, and 200–300 mg of CuO granules were placed into preheated quartz tubes for sealed quartz tube combustion at 800 °C. Quartz tubes were preheated at 800 °C for 2 h, and CuO granules were preheated at 800 °C one day before usage. Combustion was set for 2 h at 800 °C. Samples were then cooled slowly from 800 to 600 °C for 6 h to allow Cu to reduce the NxO to nitrogen gas. Purified CO_2 was collected cryogenically under vacuum conditions, which were less than 10 mTorr, and submitted to the Keck Carbon Cycle AMS Laboratory of the University of California-Irvine for AMS ^{14}C analysis using the hydrogen–iron reduction method with δ^{13}C values measured on prepared graphite [30]. All results were corrected for isotopic fractionation according to the conventions of Stuiver and Polach (1977) [31]. Sample preparation backgrounds were subtracted based on the measurements of radiocarbon-free wood blanks. The results indicated that after background subtraction, IAEA-C5 and FIRI-D wood reference materials yielded target values within 1σ deviations. Radiocarbon dating data greater than 100% of modern C were considered as present C, which was fixed from 1950 to 2012.

Soil texture was determined using the laser diffraction method [32]. Soil pH was measures in a 1:1 ratio of soil to water. Plant-available, exchangeable potassium, magnesium, and calcium were determined with the Bray-1 extraction method [33]. The content of cations, cation exchange capacity, and percent base saturation of cation elements were calculated from extract results.

2.4. Calculations

Bulk density, estimated by dividing the oven-dried mass (g) by the canister volume, C content in <2 mm bulk soil (%), and the soil bulk density (BD, g cm^{-3}), was used to estimate soil C stock per 10 cm depth increments (Mg C ha^{-1}).

$$\text{Soil C stock} = \text{Soil C content} \times BD \times 10, \tag{1}$$

To estimate C$_4$ plant contribution to soil C, $\delta^{13}C$ of soil C ($\delta^{13}C_{SC}$, ‰) were calculated as follows:

$$\delta^{13}C_{SC} = [(R_{sample}/R_{standard} - 1)] \times 1000, \tag{2}$$

where R is the ratio of $^{13}C/^{12}C$ in bulk soil C. The standard was V-PeeDee Belemnite (V-PDB) carbonate. The measured $\delta^{13}C$ values were converted to relative abundances of C$_3$ and C$_4$ plants using the mass balance equation:

$$\delta^{13}C_{SC} = \{(\delta^{13}C_{C4}) \times x\} + \{(\delta^{13}C_{C3}) \times (1 - x)\}, \tag{3}$$

where x indicates the ratios of C source derived from C$_4$ and C$_3$ plants, which were −13‰ and −27‰, respectively, and were used as the average values of $\delta^{13}C_{C4}$ and $\delta^{13}C_{C3}$ for calculation.

Linear regression, completed using PROC REG in SAS (version 9.2, Carey, NC, USA), was used to estimate soil C age at various soil depths with $p < 0.05$. The goodness of model fit and significance were estimated by R^2 and p-values. Profile summaries were calculated from the summed C stock for each treatment combination. Sampling sites were not compared, and only the catena position effect was assessed for total soil C and C$_4$-source C within each soil depth at each site (ANOVA, PROC GLM in SAS).

Accumulation of C$_4$-C (C_{flux}, Mg C ha^{-1} yr^{-1}) was calculated using soil C content (g C 100 g^{-1}), sedimentation rates (SR, cm yr^{-1}, from the surface down to 1.2 m), BD, and x, which is the C$_4$-derived C content from ^{13}C abundance in Equation (3) as follows:

$$C_{flux} = \text{Soil C content} \times SR \times BD \times x, \tag{4}$$

Combined with the known depth of each of the soil profiles and C stock data, radiocarbon dating of the profiles was used to generate age–depth models to estimate the sedimentation rate to 1.2 m for total C and C from C$_4$ plant sources as per the methods of Howlett et al. (2013). The risk exists for $\delta^{13}C$ to become less negative due to the fact of isotopic fractionation, which could introduce uncertainty in determining the contribution of C from C$_3$ and C$_4$ plants. However, degradation-induced fractionation is essentially negligible, because new additions of organic C in the mesic *Miscanthus*-grassland ecosystem generally overwhelm the oxidation of the soil organic C pool. Moreover, decomposition only enriches less than 1–2‰ for soils in dry and/or hot environments, which was not the case in our study.

3. Results

Selected soil physical and chemical properties are shown in Table 1. Values represent the averages of whole soil samples from the surface to a 1.2 m depth, because soil samples were collected at 10 cm soil depth increments and could not be presented by soil horizons. Selected soil physical and chemical properties show that the soils from the three *Miscanthus sinensis*-dominated grassland catenas were low in pH, had a texture from silt to silt loam, and had low to moderate CEC (Table 1). Low BD is typical of volcanic ash-derived soils.

Table 1. Soil physical and chemical properties (Average ± SD) underlying three *Miscanthus sinensis* grasslands in Japan (i.e., Kawatabi Field Science Center, Miyagi Prefecture (KAW); Soni Plateau, Nara Prefecture (SONI); Aso-Kuju National Park, Kumamoto Prefecture (ASO)).

Site	Depth (cm)	Clay (%)	Silt (%)	Sand (%)	Bulk Density (g cm^{-3})	pH	CEC [§] (cmol$_c$ kg^{-1})	Ex. [†] K (%)	Ex. Mg (%)	Ex. Ca (%)
KAW	0–30	6.9 ± 2.2	88.1 ± 0.9	4.0 ± 1.6	0.5 ± 0.0	4.6 ± 0.4	103 ± 118	2.0 ± 1.2	5.0 ± 4.6	14.9 ± 12.8
	30–60	7.3 ± 3.1	88.2 ± 2.5	3.5 ± 2.4	0.6 ± 0.3	4.9 ± 0.3	13 ± 8	2.6 ± 1.2	16.3 ± 13.7	35.4 ± 25.5
	60–90	7.1 ± 3.3	75.4 ± 16.7	16.6 ± 20.3	0.8 ± 0.4	5.2 ± 0.2	8 ± 8	5.7 ± 2.6	24.7 ± 10.9	46.2 ± 27.4
	90–120	6.5 ± 3.9	78.4 ± 5.6	14.2 ± 9.7	0.9 ± 0.2	5.3 ± 0.2	9 ± 7	7.3 ± 3.6	27.2 ± 15.7	41.9 ± 24.7
SONI	0–30	8.9 ± 0.8	88.6 ± 0.9	1.2 ± 0.1	0.6 ± 0.1	5.3 ± 0.1	15 ± 8	20.0 ± 10.1	15.7 ± 10.3	22.5 ± 16.2
	30–60	6.5 ± 4.1	84.8 ± 1.9	7.9 ± 5.5	0.9 ± 0.3	5.2 ± 0.1	4 ± 1	14.4 ± 8.2	28.5 ± 2.7	57.1 ± 5.4
	60–90	8.0 ± 4.6	85.3 ± 5.1	5.9 ± 0.2	1.0 ± 0.2	5.4 ± 0.2	4 ± 1	10.1 ± 2.7	31.9 ± 3.0	57.9 ± 4.7
	90–120	8.3 ± 3.8	84.2 ± 3.9	6.6 ± 5.5	1.1 ± 0.1	5.6 ± 0.1	4 ± 1	13.9 ± 13.7	28.7 ± 4.6	57.4 ± 9.1
ASO	0–30	3.4 ± 0.5	66.5 ± 2.5	29.6 ± 2.8	0.7 ± 0.1	5.8 ± 0.1	42 ± 23	2.6 ± 1.5	12.6 ± 3.9	49.4 ± 17.9
	30–60	3.4 ± 0.3	74.8 ± 0.5	21.3 ± 0.5	0.6 ± 0.1	5.8 ± 0.1	29 ± 4	4.4 ± 3.0	13.4 ± 2.1	41.9 ± 8.6
	60–90	3.1 ± 0.3	75.6 ± 5.6	20.9 ± 5.9	0.6 ± 0.1	6.0 ± 0.3	44 ± 14	4.4 ± 3.0	10.0 ± 4.0	59.4 ± 13.0
	90–120	4.0 ± 1.3	81.6 ± 7.1	13.8 ± 8.6	0.5 ± 0.1	6.1 ± 0.2	75 ± 40	4.4 ± 3.0	12.3 ± 0.6	62.7 ± 10.4

[§] Cation exchange capacity; [†] exchangeable.

Whole-profile soil C stocks across all sites ranged from 249 to 640 Mg C ha^{-1} for a 0–1.2 m depth (Figure 2). Across the study sites, the position along the catena sequence was a significant factor at KAW and SONI only. At KAW, soil C stock in the toe slope (640 Mg C ha^{-1}) was greater than in the summit (379 Mg C ha^{-1}) but statistically similar to that in the mid-slope (532 Mg C ha^{-1}). At SONI, the pattern of the distribution of soil C stock was similar as in the mid-slope (483 Mg C ha^{-1}) and in the toe slopes (358 Mg C ha^{-1}) exceeded in the summit (249 Mg C ha^{-1}). However, no differences in the soil C stock among catena positions were found at ASO.

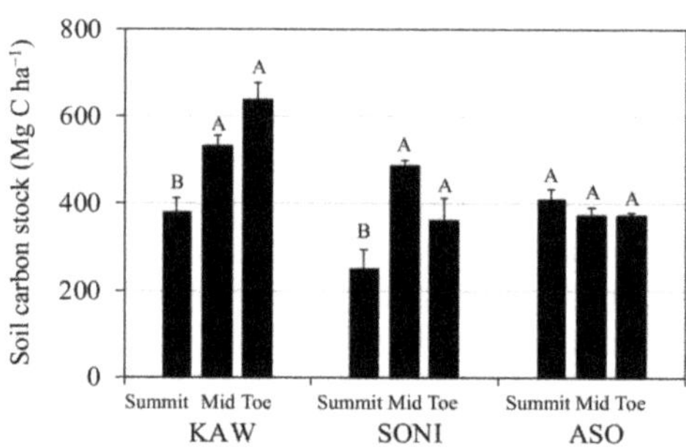

Figure 2. Profile summary of mean soil carbon stock along three catenas at semi-natural *Miscanthus sinensis* grassland sites in Japan (i.e., Kawatabi Field Science Center, Miyagi Prefecture (KAW); Soni Plateau, Nara Prefecture (SONI); Aso-Kuju National Park, Kumamoto Prefecture (ASO)). Error bars represent the standard errors. Statistically different means within a site are noted by means separation letters ($p < 0.05$).

Catena position was also a significant factor for soil C stocks at certain depths at each site (Figure 3, Supplement Table S1). At KAW, soil C stock in the toe slope was greater than that in the summit slope from 50 to 120 cm (Figure 3a), while it was higher in surface soil at the summit. At SONI, soil C stock in the mid-slopes demonstrated higher levels only between 50 and 100 cm depths, but it was statistically indistinguishable from toe slopes at most of these depths (Figure 3b). At ASO, a relatively higher soil C stock at the summit was observed from the surface down to a 20 cm depth (Figure 3). It was lower at the summit compared to those at the mid- and toe slopes from 30 to 80 cm depths of soil. Below 80 cm of soil, however, soil C stock increased and was higher at the summit.

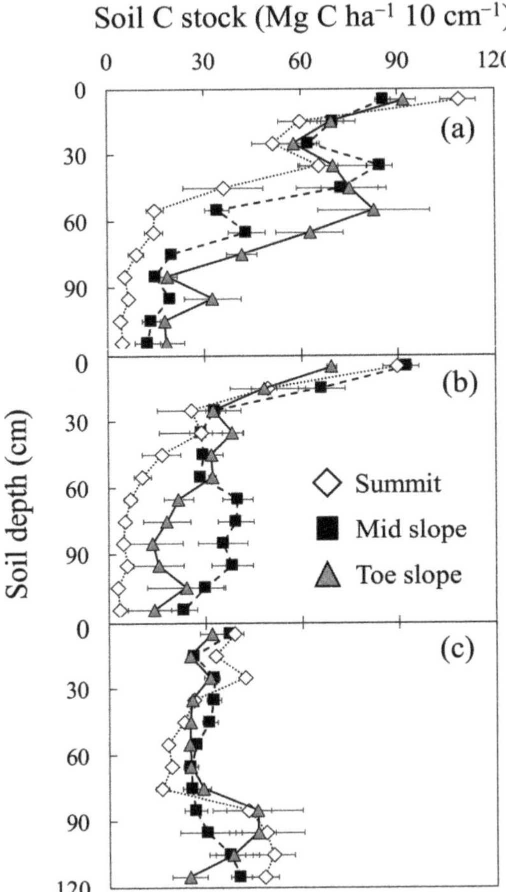

Figure 3. Soil carbon stock at 10 cm increments down to a 1.2 m depth for three positions along catenas at three semi-natural *Miscanthus sinensis* grassland sites in Japan: Kawatabi Field Science Center, Miyagi Prefecture (KAW) (**a**); Soni Plateau, Nara Prefecture (SONI) (**b**); Aso-Kuju National Park, Kumamoto Prefecture (ASO) (**c**). Error bars represent the standard errors.

The soil C accumulation rate for C_4-based C across all sites within 10 cm soil depth increments ranged from 0.00 to 0.29 Mg C ha^{-1} yr^{-1} (Figure 4). Within each site, mean C_4-C accumulation for the whole profile (0–120 cm) was 0.05 ± 0.02 (Average ± SE), 0.04 ± 0.00, and 0.24 ± 0.04 Mg C ha^{-1} yr^{-1} at KAW, SONI, and ASO, respectively. At ASO, C_4-C constituted the vast majority of total C, especially from 80 to 120 cm (Figure 4c). To a lesser extent, KAW soil C was mostly C_4-C (Figure 4a). In addition, C_4-C closely followed the trend of total C, decreasing in content from 40 to 80 cm. At SONI, C_4-C comprised the majority of total C (Figure 4b). ASO had the highest mean content of C_4-C at 86.3% (57.0–100%) with KAW at 58.2% (28.6–99.1%) and SONI at 56.3% (37.0–76.9%).

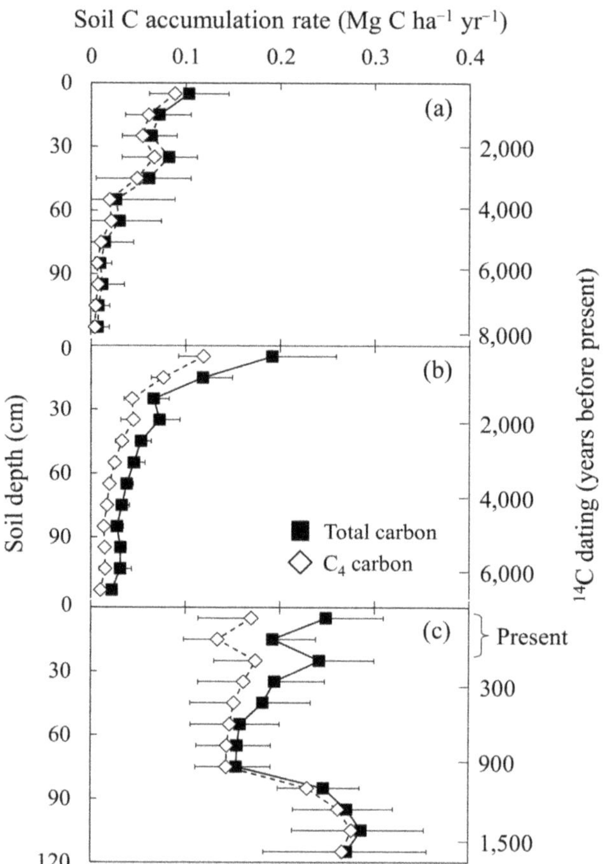

Figure 4. Soil carbon accumulation rate and relative soil age (years before present) for C$_4$-derived C (clear diamonds) and total C (black squares) to a 1.2 m in three semi-natural *Miscanthus sinensis* grassland sites in Japan: Aso-Kuju National Park, Kyushu Prefecture (ASO) (**a**); Kawatabi Field Science Center, Miyagi Prefecture (KAW) (**b**); and Soni Plateau, Nara Prefecture (SONI) (**c**). Error bars represent the standard errors.

One major difference between ASO and the two other sites was the age of the bottom soil depth at 1.2 m. At ASO, the age of the 110–120 cm depth was dated 1590 years before present, while the 110–120 cm depth at KAW and SONI was closer to 7836 and 6415 years before present, respectively (Figure 4). As such, the profiles at ASO represent a more recent portion of the age ranges found in the other sites (Figure 4). The age-to-depth models used to calibrate the soil ages throughout the profile were highly correlated with an R^2 in the range of 0.83–0.98 and $p < 0.05$ (Figure 5, Supplement Table S2). The only exception was the toe slope at SONI with an R^2 of 0.79 (Figure 5b).

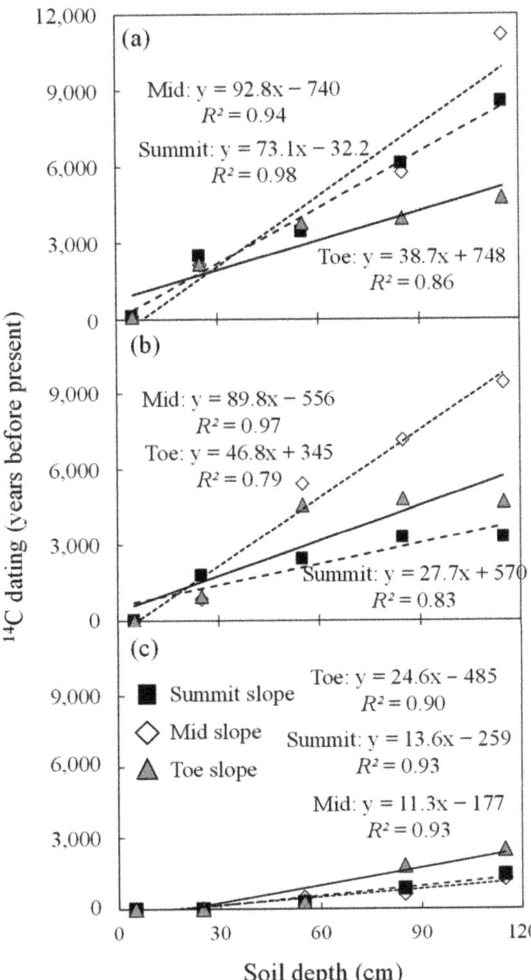

Figure 5. Calibrated radiocarbon ages to a 1.2 m depth of soil carbon at three catena positions (i.e., toe slope, mid-slope, and summit) in three semi-natural *Miscanthus sinensis* grassland sites in Japan: Kawatabi Field Science Center, Miyagi Prefecture (KAW) (**a**); Soni Plateau, Nara Prefecture (SONI) (**b**); Aso-Kuju National Park, Kyushu Prefecture (ASO) (**c**).

4. Discussion

4.1. Soil Carbon Stock along Catena

Soil C stocks in the *M. sinensis*-dominated grasslands of Japan appeared to be influenced by catena position along a toposequence. For total accumulated soil C stock for the 0–120 cm soil depths, toe and mid-slopes demonstrated a long-understood tendency to be the recipient of C-containing sediments from the summit (Figure 3) [21,22,34]. Jenny (1941) [34] indicated that erosion does not play a significant role in some well-vegetated catenas, given their minimal degree of erosion. However, the annual cultural practice of burning the *M. sinensis* grasslands in Japan reduces biotic control of erosion. Movement of sediment from organic horizons follows the course of gravity and increases C stock in mid- and toes slopes (Figure 3). This may partially help explain the pattern of soil

C accumulation found at KAW and SONI (Figure 3a,b). Upon further investigation of differences in soil C stocks at various depths within the catenas, many of the differences seen in KAW and SONI only occurred below 60 cm depths (Figure 3a,b), coinciding with soil ages in the range of approximately 4000 years before present. Differences in soil C in deeper layers across different topographic positions unlikely reflect differences in current vegetation, because *M. sinensis* rhizomes and roots mainly populated the surface layer down to a 20 cm depth [35]. Because the age of C in sediments at ASO was much younger than the other study sites, investigations of deeper soil profiles at ASO might be required to determine if catena position is a significant factor in determining soil C stocks at ASO, where no differences were found. While catena position does appear to affect soil C stock at KAW and SONI, the effect occurred between 4000 and 8000 years before present.

In all catena grasslands examined, very high total soil C stocks were found in upland (non-hydric) soils (up to 638 Mg C ha^{-1}) (Figure 4). The likely presence of high contents of Al and Fe possibly contributed to large quantities of humus stabilization in the volcanic soils of this study [11,36,37]. Formation of recalcitrant organo–mineral complexes with Al and Fe reduces translocation and mineralization of C in the soil [38,39]. This may also help explain how the relatively high amounts of sequestered C [36,40] and low pH (~5), especially at KAW, contributed to the formation of these complexes [41]. In addition, the presence of these organo–mineral complexes might possibly explain the relatively high correlations of determination that provided confidence to the sedimentation rate calculations used to estimate C accumulation. Because we could not analyze the organo–mineral complexes with Al and Fe in this study, these analyses and evaluation need to be addressed in future research. Furthermore, regularly occurring fire events over hundreds of years at ASO and SONI may also have contributed to the stabilization of soil C. Burning has been shown to increase the stability of organic matter through the formation of highly condensed aromatic compounds [42]. Thus, the evaluation of soil humus characteristics could be important variables to include in future studies.

Toma et al. (2012) [13] and Howlett et al. (2013) [11] provided a broad review of work on C sequestration in soils where *Miscanthus* has been long established. Previous work at ASO demonstrated high total C stock levels down to a 1.5 m depth (515 and 559 Mg C ha^{-1}) in two soil profiles dated, at a maximum, to 12,000 years before present [11]. The site, characterized by Howlett et al. (2013) [11], was relatively flat where erosion appeared to not be a significant factor. We considered nine soil profiles at ASO only to a depth of 1.2 m on younger soils where humus may not have had as much time to accumulate. As such, the results reported here appear to be consistent with previous work at ASO [11]. However, as with the site studied by Howlett et al. (2013) [11], there appears to be a buried organic soil horizon nearly 80 cm below the soil surface as reflected by the notable increase in soil C accumulation rates starting at that depth (Figure 4c). Basile-Doelsch et al. (2005) [43] reported high soil C stock levels 100 cm belowground of a volcanic–ash soil, which was located adjacent to the Piton des Neiges volcano on the island of La Reunion, where a burial event occurred sometime in the distant past.

4.2. Source and Rate of Carbon Accumulation

We assumed, for the purpose of this study, that all C$_4$-C was derived from *Miscanthus*, as no other known species in the study areas utilize the C$_4$ photosynthetic pathway. Miyabuchi and Sugiyama (2006) [44] detailed the dominance of *M. sinensis* via plant phytolith analysis in semi-natural grasslands located on the east side of the Aso caldera in Aso-Kuju National Park. While accumulation of C from C$_4$ sources follows the trend of total soil C accumulation throughout the profiles examined here, the content of C$_4$-C varied considerably among sites (Figure 4). Soil at ASO had the highest amount of C$_4$-derived C accumulation, representing nearly 100% of C from 70 to 120 cm (Figure 4c). The C$_4$-C and total-C accumulation trends at KAW and SONI underscore the importance of *Miscanthus*-derived C, but since not all C was from C$_4$ sources, additions of C from non-*Miscanthus* sources were consistent with the presence of other plant species over time.

Although currently dominated by *M. sinensis*, the composition of vegetative inputs in the plant community to soil C varied over the 12,000 year period [11,25]. At SONI, previous work identified charcoal remnants from anthropogenic fires that began around 7000 years before present with phytolith data indicating a vegetative shift [25]. If C_4-C was mostly *Miscanthus*-derived, the general increase in C_4-C accumulation at KAW (Figure 4a) may indicate a vegetative change from forest to grassland, which may have promoted more C storage in soil C pools relative to aboveground biomass [13,45,46].

Rates of *Miscanthus*-source soil C accumulation, highest at ASO, may be an indication of the greater net primary production that occurred under warmer and nearly double the precipitation than that observed at SONI and KAW (Figure 4). Howlett et al. (2013) [11] measured soil C accumulation at ASO on a site 14 km northwest of the current study site and found mean C_4-C accumulation rates between 0.62 and 0.85 Mg C ha^{-1} yr^{-1} down to 1.5 m in the soil. These previously studied profiles likely represented several buried organic horizons dating to approximately 12,000 years before present, where humus accumulation occurred over an extensive period. As mentioned above, a similar phenomenon appears to have occurred at the current study site, where a buried organic horizon appears nearly 80 cm below the soil surface but is considerably younger (Figure 4c).

As suggested by Chaopricha and Marin-Spiotta (2014) [47], soil burial is a globally important, yet largely underestimated, process involved in the storage and persistence of substantial C stocks in soils. Indeed, volcanic soils buried 3 m below the surface on the slopes of Mount Kilimanjaro were estimated to contain 820 Mg C ha^{-1} [48]. In addition, Inoue et al. (2000) [49] found that high soil C levels, which were buried multiple times over for several thousand years in a volcanic basin 135 km south of Aso, had not substantially decreased since the initial burial events. Similarly, based on our data and that of Howlett et al. (2013) [11], we strongly suspect that large reservoirs of C are stored in buried soils throughout the Aso volcanic caldera. Indeed, several volcanic eruptions have occurred in the ASO area over the past several thousand years, including several that occurred in the early 1200s [50], which coincide with the putative burial event seen in the soil profile at the current study site. These events suggest that soil burial due to the soil sedimentation resulting in volcanic ash deposition and plant residue accumulation acts as an important process in maintaining soil C levels at deeper soil layers under the stable thermal environments and anaerobic conditions. Possibly due to the more recent eruption event, the current study site had much more ash deposition in the subsurface soil than the study site of Howlett et al. (2013) [11], which was likely due to the current site being 6.3 km closer to the volcano at Aso. Moreover, more ash deposition likely occurred given the west-to-east prevailing wind direction in the region. The study site of Howlett et al. (2013) [11] was 15.9 km north of the volcano, whereas the current study site was 9.7 km east of the volcano. In this study, we report C_4-C accumulation rates that were 3–4 times lower than that reported by Howlett et al. (2013) [11]. Given that the soil-C measurements of the current study were taken to only a 1.2 m depth in comparatively younger soils (to 1590 years before present) at ASO may explain the lower soil C accumulation rates. Zehetner (2010) [51] reported that soil C accumulation rates can range between 0.3 and 0.6 Mg C ha^{-1} yr^{-1} in relatively volcanic–ash soils. However, most studies on soil C accumulation in cultivated *Miscanthus* fields have reported considerably higher rates. Soils where *M. sinensis* had been established for 6 [7] and 14 years [52] under managed conditions in southeastern England accumulated C at approximately 0.80 Mg ha^{-1} yr^{-1}, which is similar to what Poeplau and Don (2014) [53] found in an analysis of six *Miscanthus* plantations $\geq$10 years old across Europe (0.78 Mg ha^{-1} yr^{-1}). In addition, based on 23 data sets, Agostini et al. (2015) and Qin et al. (2016) [54] both calculated global estimates of C accumulation under *Miscanthus* to be approximately 1.2 Mg ha^{-1} yr^{-1}. Differences in soil clay content, soil bulk density, and initial low C stocks between the semi-natural *Miscanthus* grassland site and the primarily managed fields in these other studies may have led to considerable differences in soil C sequestration [55–57]. In addition, given that the managed fields were amended with fertilizer, this undoubtedly contributed to the differences in soil C sequestration rates.

KAW and SONI had mean C_4-C accumulation rates roughly an order of magnitude less than ASO. These colder, more northern latitude sites, with half the precipitation of ASO, likely have lower net primary production. As such, potential C inputs to the soil would be expected to be lower.

5. Conclusions

As *Miscanthus* becomes more widely planted outside its native range, particularly in low soil C agronomic fields, the potential exists for sequestration of C over the long term. Moreover, anthropogenic fire events, which are used to maintain vegetation, may further increase soil C. Toposequence along a catena influence soil C stocks in *M. sinensis* grasslands in its native range of Japan. Consideration of C sequestration in cultivated *Miscanthus* fields should include characterization of topographic variability. A majority of soil C in the grasslands examined appears to have derived from C_4-C. In addition, accumulation rates for C_4-C were lower than previously demonstrated.

Supplementary Materials: The following supporting information can be downloaded at: https://www.mdpi.com/article/10.3390/agriculture12010088/s1, Table S1: Calibrated radiocarbon ages to 1.2 m depth of soil carbon at three catena positions (toe slope, mid slope, and summit) in three semi-natural Miscanthus sinensis grassland sites in Japan: Kawatabi Field Science Center, Miyagi Pre-fecture (KAW), and Soni Plateau, Nara Prefecture (SONI), and Aso-Kuju National Park, Kyushu Prefecture (ASO); Table S2: Calibrated radiocarbon ages to 1.2 m depth of soil carbon at three catena positions (toe slope, mid slope, and summit) in three semi-natural Miscanthus sinensis grassland sites in Japan: Kawatabi Field Science Center, Miyagi Pre-fecture (KAW), and Soni Plateau, Nara Prefecture (SONI), and Aso-Kuju National Park, Kyushu Prefecture (ASO).

Author Contributions: Conceptualization, D.S.H., J.R.S., T.Y., F.G.F. and Y.T.; methodology, D.S.H., J.R.S., J.I., M.S., A.N., F.G.F. and Y.T.; software, D.S.H.; validation, D.S.H. and Y.T.; formal analysis, D.S.H., H.W. and Y.T.; investigation, D.S.H., J.I., M.S. and Y.T.; resources, J.R.S., D.L., H.W., T.Y. and F.G.F.; data curation, D.S.H., J.R.S. and Y.T.; writing—original draft preparation, D.S.H.; writing—review and editing, D.S.H., J.R.S., J.I., T.Y., F.G.F., M.S. and Y.T.; visualization, D.S.H. and Y.T.; supervision, J.R.S. and Y.T.; project administration, J.R.S. and T.Y.; funding acquisition, J.R.S. and T.Y. All authors have read and agreed to the published version of the manuscript.

Funding: This project was funded by the Energy Biosciences Institute at the University of Illinois through a grant from the British Petroleum Corporation.

Acknowledgments: We would like to thank Makoto Nakaboh, and his staff at the Kyushu Biomass Forum for providing assistance to our field research. We would also like to express our appreciation to the Aso Environmental Office and owners of the study site in Aso, Kumamoto, Japan; student workers at Kawatabi Field Science Center; the Geology Department of Osaka City University; Carolina Bueno Wandscheer, who provided invaluable assistance to field collections.

Conflicts of Interest: The authors declare no conflict of interest. The funders had no role in the design of the study; in the collection, analyses, or interpretation of data; in the writing of the manuscript or in the decision to publish the results.

References

1. Clifton-Brown, J.; Hastings, A.; Mos, M.; McCalmont, J.P.; Ashman, C.; Awty-Carroll, D.; Cerazy, J.; Chiang, Y.C.; Cosentiono, S.; Cracroft-Eley, W.; et al. Progress in upscaling *Miscanthus* biomass production for the European bio-economy with seed-based hybrids. *Glob. Chang. Biol. Bioenergy* **2017**, *9*, 6–17. [CrossRef]
2. Huang, L.S.R.; Flavell, R.; Donnison, I.S.; Chiang, Y.C.; Hastings, A.; Hayes, C.; Heidt, C.; Hong, H.; Hsu, T.W.; Humphreys, M.; et al. Collecting wild *Miscanthus* germplasm in Asia for crop improvement and conservation in Europe whilst adhering to the guidelines of the United Nations' Convention on Biological Diversity. *Ann. Bot.* **2018**, *124*, 591–604. [CrossRef]
3. Somerville, C.; Youngs, H.; Taylor, C.; Davis, S.C.; Long, A.P. Feedstocks for lignocellulosic biofuels. *Science* **2010**, *329*, 790–792. [CrossRef] [PubMed]
4. Stewart, J.R.; Toma, Y.; Fernández, F.G.; Nishiwaki, A.; Yamada, T.; Bollero, G. The ecology and agronomy of *Miscanthus sinensis*, a species important to bioenergy crop development, in its native range in Japan: A review. *Glob. Chang. Biol. Bioenergy* **2009**, *1*, 126–153. [CrossRef]

5. Agostini, F.; Gregory, A.S.; Richter, G.M. Carbon sequestration by perennial energy crops: Is the jury still out? *BioEnergy Res.* **2015**, *8*, 1057–1080. [CrossRef]
6. Clifton-Brown, J.C.; Breur, L.; Jones, M.B. Carbon mitigation by the energy crop. *Miscanthus. Glob. Change Biol.* **2007**, *13*, 2296–2307. [CrossRef]
7. Gregory, A.S.; Dungait, J.A.J.; Shield, I.F.; Macalpine, W.J.; Cunniff, J.; Durenkamp, M.; White, R.P.; Joynes, A.; Richter, G.M. Species and genotype effects of bioenergy crops on root production, carbon and nitrogen in temperate agricultural soil. *Bioenergy Res.* **2018**, *11*, 382–397. [CrossRef]
8. Hansen, E.M.; Christensen, B.T.; Jensen, L.S.; Kristensen, K. Carbon sequestration in soil beneath long-term *Miscanthus* plantations as determined by ^{13}C abundance. *Biomass Bioenergy* **2004**, *26*, 97–105. [CrossRef]
9. Schneckenberger, K.; Kuzyakov, Y. Carbon sequestration under *Miscanthus* in sandy and loamy soils estimated by natural ^{13}C abundance. *J. Plant Nutr. Soil Sci.* **2007**, *170*, 538–542. [CrossRef]
10. Anzoua, K.G.; Suzuki, K.; Fujita, S.; Toma, Y.; Yamada, T. Evaluation of morphological traits, winter survival and biomass potential in wild Japanese *Miscanthus sinensis* Anderss. populations in northern Japan. *Grassl. Sci.* **2015**, *61*, 83–91. [CrossRef]
11. Howlett, D.S.; Toma, Y.; Wang, H.; Sugiyama, S.; Yamada, T.; Nishiwaki, A.; Fernández, F.G.; Stewart, J.R. Soil carbon source and accumulation over 12,000 years in a semi-natural *Miscanthus sinensis* grassland in southern Japan. *Catena* **2013**, *104*, 127–135. [CrossRef]
12. Toma, Y.; Yamada, T.; Fernández, F.G.; Nishiwaki, A.; Hatano, R.; Stewart, J.R. Evaluation of greenhouse gas emissions in a *Miscanthus sinensis* Andersson-dominated semi-natural grassland in Kumamoto, Japan. *Soil Sci. Plant Nutr.* **2016**, *62*, 80–89. [CrossRef]
13. Toma, Y.; Armstrong, K.; Stewart, J.R.; Yamada, T.; Nishiwaki, A.; Fernández, F.G. Carbon sequestration in soil in a semi-natural *Miscanthus sinensis* grassland and Cryptomeria japonica forest plantation in Aso, Kumamoto, Japan. *Glob. Chang. Biol. Bioenergy* **2012**, *4*, 566–575. [CrossRef]
14. Toma, Y.; Fernández, F.G.; Sato, S.; Izumi, M.; Hatano, R.; Yamada, T.; Nishiwaki, A.; Bollero, G.; Stewart, J.R. Carbon budget and methane and nitrous oxide emissions over the growing season in a *Miscanthus sinensis* grassland in Tomakomai, Hokkaido, Japan. *Glob. Chang. Biol. Bioenergy* **2011**, *3*, 116–134. [CrossRef]
15. Toma, Y.; Fernández, F.G.; Nishiwaki, A.; Yamada, T.; Bollero, G.; Stewart, J.R. Aboveground plant biomass, carbon, and nitrogen dynamics before and after burning in a seminatural grassland of *Miscanthus sinensis* in Kumamoto, Japan. *Glob. Chang. Biol. Bioenergy* **2010**, *2*, 52–62. [CrossRef]
16. Boutton, T.W.; Archer, S.R.; Midwood, A.J.; Zitzer, S.F.; Bol, R. δ^{13}C values of soil organic carbon and their use in documenting vegetation change in a subtropical savanna ecosystem. *Geoderma* **1998**, *82*, 5–41. [CrossRef]
17. Farquhar, G.; Ehleringer, J.; Hubick, K. Carbon isotope discrimination and photosynthesis. *Annu. Rev. Plant Physiol. Plant Mol. Biol.* **1989**, *40*, 503–537. [CrossRef]
18. Fry, B. *Stable Isotope Ecology*, 1st ed.; Springer Science & Business Media: New York, NY, USA, 2007.
19. Wang, H.; Hackley, K.C.; Panno, S.V.; Coleman, D.D.; Liu, J.C.L.; Brown, J. Pyrolysis-combustion ^{14}C dating of soil organic matter. *Quat. Res.* **2003**, *60*, 348–355. [CrossRef]
20. Wang, Y.; Amundson, R.; Trumbore, S. Radiocarbon dating of soil organic matter. *Quat. Res.* **1996**, *45*, 282–288. [CrossRef]
21. Berhe, A.A.; Barnes, R.T.; Six, J.; Marin-Spiotta, E. Role of soil erosion in biogeochemical cycling of essential elements: Carbon, nitrogen, and phosphorus. *Annu. Rev. Earth Planet. Sci.* **2018**, *46*, 521–548. [CrossRef]
22. Schimel, D.; Stillwell, M.A.; Woodmansee, R.G. Biogeochemistry of C, N, and P in a soil catena of the shortgrass steppe. *Ecology* **1985**, *66*, 276–282. [CrossRef]
23. Ito, T.; Saigusa, M. Characteristics of nonallophanic andisols at Tohoku University Farm. *Bull. Exp. Farm Tohoku Univ.* **1996**, *12*, 91–103, (In Japanese with English Summary).
24. Inoue, J.; Nishimura, R.; Takahara, H. A 7500-year history of intentional fires and changing vegetation on the Soni Plateau, Central Japan, reconstructed from macroscopic charcoal and pollen records within mire sediment. *Quat. Int.* **2012**, *254*, 12–17. [CrossRef]
25. Okunaka, R.; Kawano, T.; Inoue, J. Holocene history of intentional fires and grassland development on the Soni Plateau, Central Japan, reconstructed from phytolith and macroscopic charcoal records within cumulative soils, combined with paleoenvironmental data from mire sediments. *Holocene* **2012**, *22*, 793–800. [CrossRef]
26. Miyabuchi, Y.; Sugiyama, S.; Nagaoka, Y. Vegetation and fire history during the last 30,000 years based on phytolith and macroscopic charcoal records in the eastern and western areas of Aso Volcano, Japan. *Quat. Int.* **2012**, *254*, 28–35. [CrossRef]
27. USDA NRCS. *Keys to Soil Taxonomy*; United States Department of Agriculture: Washington, DC, USA, 2014.
28. Miyabuchi, Y. A 90,000-year tephrostratigraphic framework of Aso Volcano, Japan. *Sediment. Geol.* **2009**, *220*, 169–189. [CrossRef]
29. Brandt, S.A.; Fisher, F.C.; Hildebrand, E.A.; Vogelsang, R.; Ambrose, S.H.; Lesur, J.; Wang, H. Early MIS 3 occupation of Mochena Borago Rockshelter, Southwest Ethiopian Highlands: Implications for Late Pleistocene archaeology, paleoenvironments and modern human dispersals. *Quat. Int.* **2012**, *274*, 38–54. [CrossRef]
30. Southon, J. Graphite reactor memory—Where is it from and how to minimize it? *Nucl. Instrum. Methods Phys. Res. Sect. B Beam Interact. Mater. Atoms* **2007**, *259*, 288–292. [CrossRef]
31. Stuiver, M.; Polach, H.A. Discussion reporting of 14C Data. *Radiocarbon* **1977**, *19*, 355–363. [CrossRef]
32. Eshel, G.; Levy, G.J.; Mingelgrin, U.; Singer, M.J. Critical evaluation of the use of laser diffraction for particle-size distribution analysis. *Soil Sci. Soc. Am. J.* **2004**, *68*, 736–743. [CrossRef]

33. Bray, R.H.; Kurtz, L.T. Determination of total, organic, and available forms of phosphorus in soils. *Soil Sci.* **1945**, *59*, 39–45. [CrossRef]

34. Jenny, H. *Factors of Soil Formation: A System of Quantitative Pedology*; McGraw-Hill: New York, NY, USA, 1942.

35. Yano, N. On the subterranean organ of wild plants of grassland, I. Miscanthus sinensis Andresson. *Jpn. J. Grassl. Sci.* **1965**, *11*, 48–54, (In Japanese with English Summary). [CrossRef]

36. Dahlgren, R.A.; Saigusa, M.; Ugolini, F.C. *Advances in Agronomy: The Nature, Properties and Management of Volcanic Soils*; Elsevier Academic Press Inc.: San Diego, CA, USA, 2004.

37. Dahlgren, R.A.; Ugolini, F.C.; Shoji, S.; Ito, T.; Sletten, R.S. Soil-forming processes in Alic Melanudands under Japanese pampas grass and oak. *Soil Sci. Soc. Am. J.* **1991**, *55*, 1049–1056. [CrossRef]

38. Matus, F.; Rumpel, C.; Neculman, R.; Panichini, M.; Mora, M.L. Soil carbon storage and stabilisation in andic soils: A review. *Catena* **2014**, *120*, 102–110. [CrossRef]

39. Pena-Ramirez, V.M.; Vazquez-Selem, L.; Siebe, C. Soil organic carbon stocks and forest productivity in volcanic ash soils of different age (1835–30,500 years BP) in Mexico. *Geoderma* **2009**, *149*, 224–234. [CrossRef]

40. Boudot, J.P.; Hadj, B.A.B.; Chone, T. Carbon mineralization in andosols and aluminium-rich highland soils. *Soil Biol. Biochem.* **1986**, *18*, 457–461. [CrossRef]

41. Shoji, S.; Kurebayashi, T.; Yamada, I. Growth and chemical composition of Japanese pampas grass (*Miscanthus sinensis*) with special reference to the formation of dark-colored andisols in northeastern Japan. *Soil Sci. Plant Nutr.* **1990**, *36*, 105–120. [CrossRef]

42. Johnson, W.C. Sequestration in buried soils. *Nat. Geosci.* **2014**, *7*, 398–399. [CrossRef]

43. Basile-Doelsch, I.; Amundson, R.; Stone, W.E.E.; Masiello, C.A.; Bottero, J.Y.; Colin, F.; Masin, F.; Borschneck, D.; Meunier, J.D. Mineralogical control of organic carbon dynamics in a volcanic ash soil on La Reunion. *Eur. J. Soil Sci.* **2005**, *56*, 689–703. [CrossRef]

44. Miyabuchi, Y.; Sugiyama, S. A 30,000-year phytolith record of a tephra sequence, east of Aso caldera, Southwestern Japan. *Quant. Res.* **2006**, *45*, 15–28, (In Japanese with English Summary). [CrossRef]

45. Guo, L.B.; Gifford, R.M. Soil carbon stocks and land use change: A meta analysis. *Glob. Change Biol.* **2002**, *8*, 345–360. [CrossRef]

46. Howlett, D.S.; Mosquera-Losada, M.R.; Nair, P.K.R.; Nair, V.D.; Rigueiro-Rodríguez, A. Soil carbon storage in silvopastoral systems and a treeless pasture in northwestern Spain. *J. Environ. Qual.* **2011**, *40*, 825–832. [CrossRef]

47. Chaopricha, N.T.; Marin-Spiotta, E. Soil burial contributes to deep soil organic carbon storage. *Soil Biol. Biochem.* **2014**, *69*, 251–264. [CrossRef]

48. Zech, M.; Horold, C.; Leiber-Sauheitl, K.; Kuhnel, A.; Andreas, H.; Zech, W. Buried black soils on the slopes of Mt. Kilimanjaro as a regional carbon storage hotspot. *Catena* **2014**, *112*, 125–130. [CrossRef]

49. Inoue, Y.; Sugiyama, S.; Nagatomo, Y. Organic carbon content and phytolith in a cumulative Andisols in Miyakonojo Basin, Japan. *Pedologist* **2000**, *44*, 109–123, (In Japanese with English Summary). [CrossRef]

50. Smithsonian Institution Asosan. Available online: https://volcano.si.edu/volcano.cfm?vn=282110 (accessed on 14 November 2021).

51. Zehetner, F. Does organic carbon sequestration in volcanic soils offset volcanic CO_2 emissions? *Quat. Sci. Rev.* **2010**, *29*, 1313–1316. [CrossRef]

52. Richter, G.M.; Agostini, F.; Redmile-Gordon, M.; White, R.; Goulding, K.W.T. Sequestration of C in soils under Miscanthus can be marginal and is affected by genotype-specific root distribution. *Agric. Ecosyst. Environ.* **2015**, *200*, 169–177. [CrossRef]

53. Poeplau, C.; Don, A. Soil carbon changes under *Miscanthus* driven by C_4 accumulation and C_3 decompostion—Toward a default sequestration function. *Glob. Chang. Biol. Bioenergy* **2014**, *6*, 327–338. [CrossRef]

54. Qin, Z.C.; Dunn, J.B.; Kwon, H.Y.; Mueller, S.; Wander, M.M. Soil carbon sequestration and land use change associated with biofuel production: Empirical evidence. *Glob. Chang. Biol. Bioenergy* **2016**, *8*, 66–80. [CrossRef]

55. Campbell, J.E.; Lobell, D.B.; Genova, R.C.; Field, C.B. The global potential of bioenergy on abandoned agriculture lands. *Environ. Sci. Technol.* **2008**, *42*, 5791–5794. [CrossRef]

56. Six, J.; Conant, R.T.; Paul, E.A.; Paustian, K. Stabilization mechanisms of soil organic matter: Implications for C-saturation of soils. *Plant Soil* **2002**, *241*, 155–176. [CrossRef]

57. Six, J.; Elliott, E.T.; Paustian, K. Soil macroaggregate turnover and microaggregate formation: A mechanism for C sequestration under no-tillage agriculture. *Soil Biol. Biochem.* **2000**, *32*, 2099–2103. [CrossRef]

Article

Cover Cropping Impacts Soil Microbial Communities and Functions in Mango Orchards

Zhiyuan Wei [1,2], Quanchao Zeng [1] and Wenfeng Tan [1,*]

[1] College of Resources and Environment, Huazhong Agricultural University, Wuhan 430070, China; weizhiyuan@163.com (Z.W.); quanchao@mail.hzau.edu.cn (Q.Z.)
[2] Tropical Crops Genetic Resources Institute, Chinese Academy of Tropical Agricultural Sciences, Haikou 571101, China
* Correspondence: tanwf@mail.hzau.edu.cn; Tel.: +86-27-8728-7508

Abstract: Soil microbes play critical roles in nutrient cycling, net primary production, food safety, and climate change in terrestrial ecosystems, yet their responses to cover cropping in agroforestry ecosystems remain unknown. Here, we conducted a field experiment to assess how changes in cover cropping with sown grass strips affect the fruit yields and quality, community composition, and diversity of soil microbial taxa in a mango orchard. The results showed that two-year cover cropping increased mango fruit yields and the contents of soluble solids. Cover cropping enhanced soil fungal diversity rather than soil bacterial diversity. Although cover cropping had no significant effects on soil bacterial diversity, it significantly influenced soil bacterial community compositions. These variations in the structures of soil fungal and bacterial communities were largely driven by soil nitrogen, which positively or negatively affected the relative abundance of both bacterial and fungal taxa. Cover cropping also altered fungal guilds, which enhanced the proportion of pathotrophic fungi and decreased saprotrophic fungi. The increase in fungal diversity and alterations in fungal guilds might be the main factors to consider for increasing mango fruit yields and quality. Our results indicate that cover cropping affects mango fruit yields and quality via alterations in soil fungal diversity, which bridges a critical gap in our understanding of the linkages between soil biodiversity and fruit quality in response to cover cropping in orchard ecosystems.

Keywords: soil microbes; cover cropping; mango orchards; sown grass; fungal diversity

Citation: Wei, Z.; Zeng, Q.; Tan, W. Cover Cropping Impacts Soil Microbial Communities and Functions in Mango Orchards. *Agriculture* **2021**, *11*, 343. https://doi.org/10.3390/agriculture11040343

Academic Editor: Yinglong Chen, Masanori Saito and Etelvino Henrique Novotny

Received: 2 March 2021
Accepted: 27 March 2021
Published: 12 April 2021

Publisher's Note: MDPI stays neutral with regard to jurisdictional claims in published maps and institutional affiliations.

1. Introduction

Cover cropping (i.e., sown grass strips) has been used as an important and effective method to improve soil fertilizer and soil carbon stacks [1]. A 12-year field experiment indicated that cover cropping increased soil organic carbon (SOC) stocks [1]. A meta-analysis showed that cover cropping contributed to the changes in global cropland soil carbon, with an overall mean change of 15.5% [2]. Compared with monospecies cover crops, cover crop mixtures sequestered more SOC [2]. Elevated SOC is also associated with improved soil health and fertility; therefore, increasing SOC may help to enhance agricultural productivity [3]. However, not all studies found that cover cropping results in SOC accumulation. Some studies demonstrated that the introduction of cover crops resulted in losses of SOC due to the faster growth of cover crops [4]. In addition to an increased carbon input, cover crops have been shown to increase biodiversity [5].

Soil harbors a rich diversity of invertebrate and microbial life, which drives biogeochemical processes at local and global scales. Soil microbes play critical roles in the nutrient cycling, climate regulation, decomposition and turnover of soil organic matter [6]. Cover cropping alters soil quality and thereby influences soil microbial communities in agro-ecosystems. The potential of cover crops to increase soil biodiversity and specific microbial patterns has been highlighted in very few studies, especially in fruit orchards. For example, sowing plant seed mixtures promoted the growth of the bacterial community and sarophytic fungi [7]. The long-term effects of green manure amendment are altered soil microbial properties, first found in a field experiment carried out in 1956 [8]. The soil microbiome plays important roles in fruit quality and production, as rhizosphere microbiome maintains plant health and primary productivity [9]. The plant-associated microbial community has been considered as the second genome of the plant, which is critical for plant health. Therefore, the cover cropping in orchards enhances the complexity and diversity of soil microbes in intensive agriculture soils, which in turn strongly influences plant health and net primary production.

The mango (*Mangifera indica* L.) is the most important fruit crop in the tropical zones, having socio-economic significance, originating from South East Asia and cultivated worldwide [10]. It is known as the king of fruits owing to its delicious taste and high vitamin C and mineral contents [11]. Mango fruit has become the second tropical crop in terms of production and cultivated acreage [12]. To steadily increase fruit yield, farmers have increased the use of chemical fertilizers, which has caused many environmental problems, such as soil degradation and water contamination [13]. Cover cropping began to be used in orchards to resolve degraded soils and maintain higher crop yields and quality [14]. This green manure method has been widely used in the fruit ecosystem worldwide. The positive effects of the long-term application of green manure have been reported by many studies [15,16]. For example, some studies found grass cover to affect SOC [17–19]. Soil carbon for plantings of switchgrass, no-till corn, and sweetgum with cover crops between the rows increased over the first 3 years [20]. However, the variations in the structures of the microbial community and diversity in soils caused by cover crops remain unknown, which have been considered as the main drivers of multiple soil functions and plant health. Therefore, the aims of this study are to identify (1) the effects of different sown grass types in a mango orchard on the diversity of soil bacteria and fungi; (2) the changes in the main taxa in response to sown grass trips; (3) the driving factors of the variations in soil microbial diversity and community structure; and (4) the link between soil microbial diversity, mango yield and fruit quality.

2. Materials and Methods

2.1. Site Description and Sampling

Mango is among the most important fruits in Hainan Province, Southern China. The planted area of this fruit is more than 25,000 ha, and the yields are more than 560,000 kg per year. The field experiment was conducted at the mango base in Tianya District, Sanya City, Hainan Province, China (109°24′ 52.70″ (E) and 18°19′54.10″ (N)). It belongs to the tropical maritime monsoon climate zone, with an annual average temperature of 25.7 °C. The highest temperature is in June, with an average of 28.7 °C. The lowest temperature is in January, with an average of 21.4 °C. The annual duration of sunshine is 2534 h. The annual average precipitation is 1347.5 mm. The terrain is a gentle area in the lower hills, and soil type is latosol [21].

A total of four treatments with different types of cover cropping were implemented, including control (no-till + herbicide, M), planting *Stylosanthes guianensis* (Z), planting *Brachiaria eruciformis* (B) between rows, and planting *Butterfly pea* (H). Each treatment was repeated three times, with a plot area of 120 m^2 (10 × 12 m) (Figure 1). All the plots were arranged in a randomized block design and separated by five 5 m buffer zones. The cover cropping was carried out in 2017. All treatments were transplanted with seedlings. Two rows of grass were planted between each mango row. The tested mango was Hongjinlong, with an age of 13 years. The row space between mango trees was 5 m. Fertilizers were applied in accordance with the conventional fertilization of fruit farmers, and no fertilizer was applied between rows.

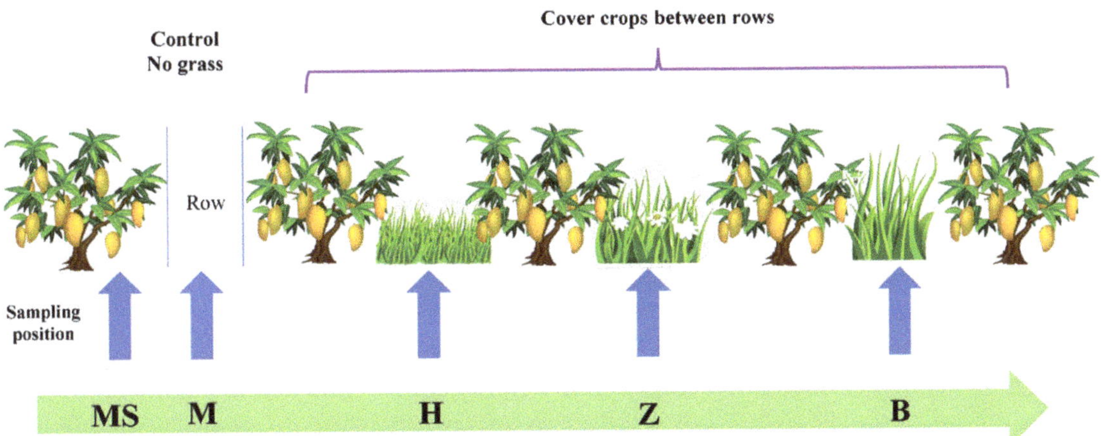

Figure 1. Experimental plot and soil sampling protocol: B, *Brachiaria eruciformis*; Z, *Stylosanthes guianensis*; H, *Butterfly pea*; M, no-till + herbicide. In the control, soils were collected between the rows (**M**) and the drip line of mango trees (**MS**). Other soils (**H**, **Z** and **B**) with cover crops were collected between the rows.

Soils were collected in July (summer) 2019. In the control, soils were collected between the rows (M) and the drip line of mango trees (MS). Other soils (H, Z and B) with cover crops were collected between the rows. In each experimental plot, we collected 20 surface soil cores (0–20 cm and 20–40 cm) randomly between rows (Figure 1). All the soil samples in one experimental plot were mixed. In total, 18 samples per soil layer were collected. After removing the roots, stones and litters by hand, we sieved them through a 2 mm sieve and separated them into two portions. One portion (approximately 5 g) was preserved in a 10 mL centrifuge tube (sterilized) and stored at −80 °C. The second portion was air-dried and used to measure soil properties.

2.2. Mango Fruit Quality and Yield

When mango fruits were nearly matured (80%), all the mango fruits in each experimental plot were harvested, and the masses were measured as yields. We randomly chose 20 mango fruits in each experimental plot and placed them into plastic bags for analysis of the fruit quality. Total soluble solids content of mango fruit was measured by a digital refractometer and presented in percent Brix (%TSS). Organic acid concentration was measured by an automatic titrator [22]. Vitamin C (Vc) content was determined via the method of 2,6-dichlorophenol indophenol (Shao et al. 2013).

2.3. Analysis of Soil Properties

Soil properties were measured by the standardized methods as described previously. Soil bulk density (BD) was measured by the volume–mass relationship via a cutting ring [23]. Soil pH was determined by a glass pH meter in a 1:2.5 soil–water suspension. Soil organic matter (SOM) was digested with 5 mL concentrated H_2SO_4 and 5 mL 0.8 M $K_2Cr_2O_7$, and then determined by 0.2 M Ferrous Ammonium Sulfate [24]. Soil total nitrogen (TN) was determined as described previously [25]. Soil available nitrogen (AVN) was extracted by 1 mol/L KCl and then measured by a Seal Auto Analyzer3 [26]. Soil total phosphorus (TP) and soil available P (AVP) were determined by molybdenum, antimony and scandium colorimetry. Total potassium (TK) and available potassium (AVK) in soil were determined by a flame photometer.

2.4. Molecular Analysis

We used the PowerSoil kit to extract DNA from 0.5 g soils. The protocol was listed in the manufacturer's instructions. After extraction, the ratios of A260/A230 and A260/A280 were determined to assess the quality of extracted DNA.

The soil bacterial community was determined by sequencing hypervariable V3-V4 regions of 16S rRNA genes using primers 338F (ACTCCTACGGGAGGCAGCA) and 806R (GGACTACHVGGGTWTCTAAT) [27]. The PCR amplification conditions for 16S rRNA were 50 s at 94 °C, 30 s at 40 °C, 35 cycles of 60 s at 72 °C, followed by 5 min at 72 °C [28]. The ITS1 variable region was sequenced with primer sets ITS3 (5′-GCATCGATGAAGAACG-CAGC-3′) and ITS4 (5′-TCCTCCGCTTATTGATATGC-3′) [29,30] to assess the soil fungal community. As for the ITS1 variable region, the PCR program was 5 min at 94 °C, 32 cycles of 30 s at 94 °C, 30 s at 54 °C, and 1 min at 72 °C [31]. Sequencing was conducted on an Illumina MiSeq PE300 platform.

After sequencing, bioinformatics processing was performed using QIIME, USEARCH and UNIOISE3 [32]. After removing the short sequences (<20 nucleotides) and chimeric sequences, the remaining sequences were clustered into operational taxonomic units (OTUs) using 97% similarity [33] (version 7.1 http://drive5.com/uparse/). Taxonomy was assigned against the Greengenes (16S gene, Release 13.5 http://greengenes.secondgenome.com/) and UNITE (ITS gene, Release 7.2 http://unite.ut.ee/index.php) databases. The Shannon diversity index and richness (the numbers of OTUs) were calculated to express the soil bacterial and fungal diversity.

2.5. Fungal Ecological Guilds Identification

We used FUNGuild to identify the fungal ecological guilds using fungal OTU dataset with taxon assignments [34]. This prediction method has been widely used in gaining insights into the distributions of soil fungal ecological groups [35,36]. In this study, we used the data with confidence levels of "highly probable" and "probable" to perform further analysis.

2.6. Statistical Analysis

We first used ANOVA to compare the differences of soil properties, mango fruit yield and quality, and bacterial and fungal diversity among different treatments on SPSS 20.0 (IBM Corporation, Armonk, NY, USA). We then conducted Pearson correlation analyses between the microbial diversity (Shannon diversity index) and soil properties. Prior to ANOVA and Pearson correlation analyses, the data were used to conduct a log transformation to meet the normality and homogeneity. Multiple regression models were constructed to compare the effects of soil properties on microbial diversity in R 3.5. Anosim analysis was performed to compare the effects of sown grass strips on the soil microbial community structure in R 3.5. The bacterial and fungal community structure was calculated based on the Bray–Curtis dissimilarity and visualized by a nonmetric multidimensional scaling (NMDS) plot in R 3.5 with the vegan package [37]. The effects of soil properties on the bacterial and fungal community structure were analyzed by the Mantel test in R 3.5. The associations between the soil microbial community structure and soil properties were determined by the Mantel test in R 3.5 [38].

3. Results

3.1. Soil Properties of the Tested Orchard

The impacts of cover cropping with sown grass strips on soil properties depended on the types of sown grass and soil layers. The upper soils (0–20 cm) were more sensitive to the application of sown grass strips. B20 had the highest soil AVN, which was significantly higher than other sown grass types. The application of sown grass strips had no significant effects on soil pH, SOM and AVK at both soil layers (Table 1).

Table 1. The characteristics of soil properties of mango soils under different sown grass strips.

Treatment	Soil layers	BD g/cm^3	Moisture %	TK%	TN%	TP%	pH	SOM %	AVP mg/kg	AVK mg/kg	AVN mg/kg
B20		1.48 ± 0.06	6.49 ± 1.19	3.44 ± 0.27	0.07 ± 0.01	0.01 ± 0	5.76 ± 0.14	1.35 ± 0.2	5.7 ± 1.59	97.43 ± 16.37	49.47 ± 7.68
H20		1.48 ± 0.01	6.02 ± 0.14	3.26 ± 0.53	0.07 ± 0	0.02 ± 0	5.55 ± 0.06	1.12 ± 0.01	6.32 ± 1.5	90.33 ± 11.77	38.27 ± 2.14
Z20	0–20 cm	1.45 ± 0.06	5.86 ± 0.66	3.49 ± 0.05	0.06 ± 0.01	0.02 ± 0	5.51 ± 0.09	1.36 ± 0.06	11.95 ± 6.12	81.07 ± 22.86	37.45 ± 5.85
M20		1.51 ± 0.04	3.75 ± 0.8	2.25 ± 0.13	0.07 ± 0	0.01 ± 0	5.49 ± 0.04	1.28 ± 0.18	7.15 ± 2.15	101.33 ± 28.45	34.07 ± 7.01
MS20		1.38 ± 0.06	5.01 ± 1.33	2.07 ± 0.3	0.07 ± 0.01	0.05 ± 0.03	5.55 ± 0.36	1.44 ± 0.11	30.03 ± 18.81	104.88 ± 11.55	51.57 ± 5.06
B40		1.46 ± 0.05	8.31 ± 1.69	3.58 ± 0.19	0.05 ± 0.01	0.02 ± 0	5.71 ± 0.17	1.16 ± 0.16	6.56 ± 0.3	93.18 ± 21.85	34.88 ± 6.95
H40		1.53 ± 0.13	7.55 ± 1.58	2.9 ± 0.53	0.06 ± 0.01	0.02 ± 0.01	5.48 ± 0.13	0.96 ± 0.04	5.3 ± 0.73	91 ± 6.58	40.6 ± 9.15
Z40	20–40 cm	1.55 ± 0.01	7.99 ± 0.8	3.32 ± 0.13	0.05 ± 0	0.02 ± 0	5.44 ± 0.14	1.19 ± 0.31	11.66 ± 5.5	86.07 ± 37.41	35.7 ± 1.4
M40		1.5 ± 0.02	7.56 ± 1.37	2.35 ± 0.17	0.05 ± 0	0.02 ± 0	5.37 ± 0.02	1.06 ± 0.27	5.57 ± 2.1	76.23 ± 8.73	29.63 ± 2.65
MS40		1.4 ± 0.04	7.08 ± 1.08	2.56 ± 0.31	0.07 ± 0.02	0.03 ± 0.01	5.4 ± 0.17	1.19 ± 0.22	42.71 ± 49.83	122.1 ± 45.21	46.43 ± 2.46

BD, soil bulk density; TK, total potassium; TN, total nitrogen; TP, total phosphorus; SOM, soil organic matter; AVP, available phosphorus; AVK, available potassium; AVN, available nitrogen.

3.2. Yield and Fruit Quality of Mango under Different Sown Grass Trips

The application of sown grass trips had significant effects on mango fruit yield, with the highest yield in B (Table 2). There were no significant differences observed between Z and M. The applications of B, H and Z significantly enhanced TSS content and decreased organic acid compared to M.

Table 2. The yield and fruit quality of mango under different sown grass trips.

Treatment	Yield (kg/ha)	TSS (%)	Vc (mg/100 g)	Organic Acid (g/kg)
B	18,736 ± 202 a	12.7 ± 0.16 a	27.97 ± 1.22 a	2.44 ± 0.13 b
H	17,424 ± 362 b	11.47 ± 0.16 b	26.31 ± 0.62 ab	2.33 ± 0.07 b
Z	16,488 ± 298 c	11.59 ± 0.3 b	24.18 ± 1.93 b	2.51 ± 0.11 ab
M	16,436 ± 189 c	10.75 ± 0.21 c	24.19 ± 0.49 b	2.93 ± 0.06 a

TSS, total soluble solids content; Vc, vitamin C. Different letters indicate significant differences under different sown grass trips. B, *Brachiaria eruciformis*; Z, *Stylosanthes guianensis*; H, *Butterfly pea*; M, no-till + herbicide. Different letter indicates significant differences between different sown grass trips.

3.3. Variations in Soil Microbial Diversity under Different Sown Grass Strips

The observed bacterial Shannon diversity index ranged from 5.47 to 6.64, while phylotype richness (OTUs) varied from 1886 to 3509. Sown grass strips have no significant effects on soil bacterial α-diversity indices (Shannon diversity index and richness index). The soil fungal phylotype richness ranged from 419 to 1080, and the fungal Shannon diversity index ranged from 3.81 to 4.95. Among these soil properties, TN and AVN had significant associations with the bacterial Shannon diversity index (Figure 2). A multiple regression models indicated that soil TN and AVN were the best predictors of the soil bacterial Shannon diversity index (with a relative importance of 0.54 and 0.42), followed by AVP, with a relative importance of 0.04. Soil properties had no significant correlations with the fungal diversity index.

3.4. Variations in Soil Microbial Community under Different Sown Grass Strips

Across all soils, a total of 1,506,304 quality bacterial sequences, and an average of 50,210 sequences per sample, were obtained. Four of the 28 phyla detected were dominant, including Acidobacteria (15.51%), Actinobacteria (17.55%), Proteobacteria (24.26%) and Chloroflexi (24.77%) (with an average relative abundance of >5%, n = 30), accounting for more than 82% of the bacterial sequences (Figure 3A). The Proteobacteria taxa were dominated by Alphaproteobacteria (17.66%), followed by Deltaproteobacteria (4.12%) and Gammaproteobacteria (2.49%). The soil layer had no significant effects on the relative abundance of Acidobacteria and Proteobacteria. The upper soils (20.5%) had a higher relative abundance of Actinobacteria than that of the lower soils (14.6%).

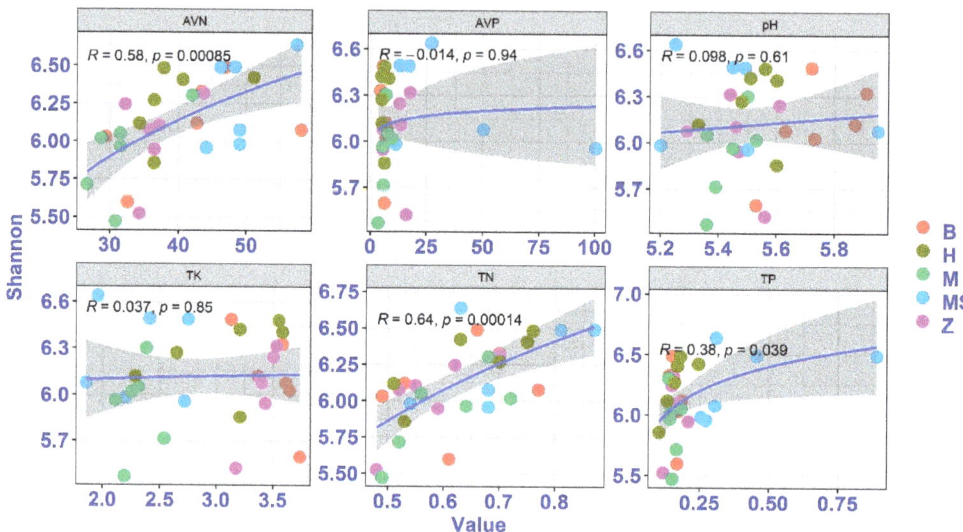

Figure 2. The associations between soil bacterial diversity and soil properties. TK, total potassium; TN, total nitrogen; TP, total phosphorus; AVP, available phosphorus; AVK, available potassium; AVN, available nitrogen. B, *Brachiaria eruciformis*; Z, *Stylosanthes guianensis*; H, *Butterfly pea*; M and MS, no-till + herbicide.

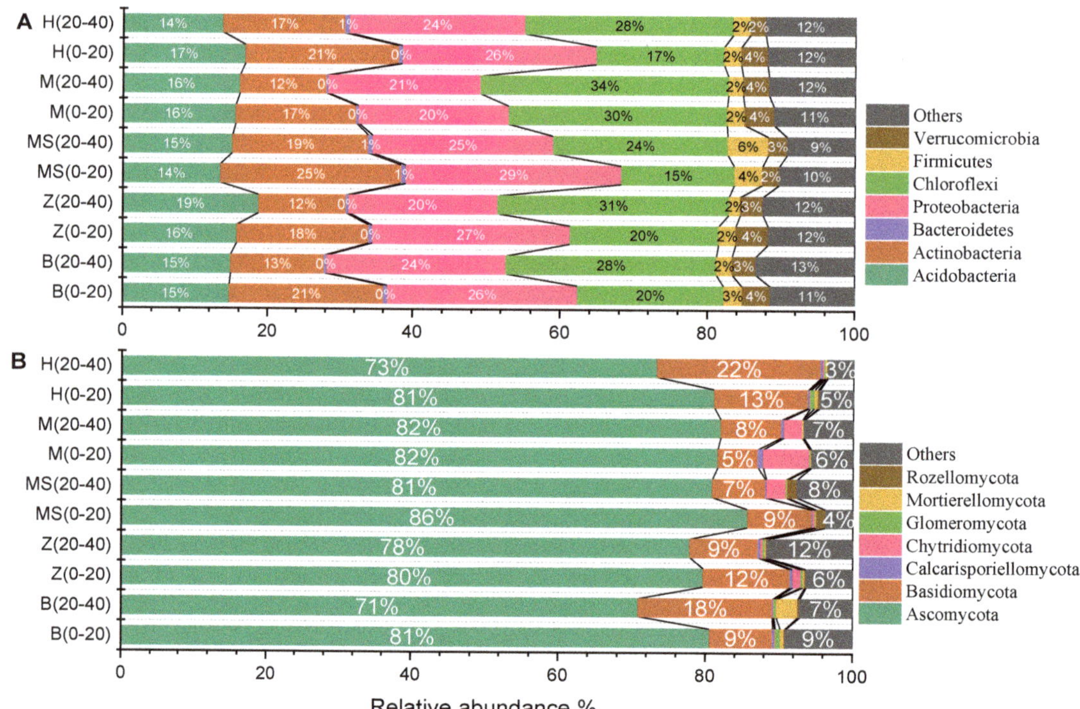

Figure 3. The community compositions of soil bacteria (**A**) and fungi (**B**) at phylum level. B, *Brachiaria eruciformis*; Z, *Stylosanthes guianensis*; H, *Butterfly pea*; M and MS, no-till + herbicide.

Across all mango soils, a total of 1,777,037 quality fungal sequences, and an average of 59,234 sequences per sample, were obtained. The dominant fungal phyla in soils were Ascomycota and Basidiomycota, with average relative abundances of 79.38 and 11.28%, respectively (Figure 3B). Other minor phyla (Anthophyta, Cercozoa, Rozellomycota and Glomeromycota) were also found at a lower relative abundance (relative abundance < 1%). The soil layer and sown grass trip had no significant effects on the relative abundance of Ascomycota and Basidiomycota. Based on taxonomical classification at the class level, Sordariomycetes (36.69%), Eurotiomycetes (18.96%), Dothideomycetes (15.33%) and Agaricomycetes (9.50%) were more abundant than other groups (relative abundance > 1%), which accounted for 80.48% of the fungal sequences. Other fungal classes were less abundant in all the soils.

Anosim analysis indicated that the fungal community structure was strongly impacted by the application of sown grass strips ($r^2 = 0.825$, $p = 0.001$ for 0–20 cm soil layer; $r^2 = 0.413$, $p = 0.001$ for 20–40 cm soil layer). Soil bacterial community structure was not sensitive to sown grass strips ($r^2 = 0.179$, $p = 0.095$ for 0–20 cm soil layer; $r^2 = 0.092$, $p = 0.203$ for 20–40 cm soil layer). Different soil samples were clearly separated by the sown grass types in the NMDS plot (Figure 4A,B). The Mantel test showed that SOM ($r = 0.16$, $p = 0.036$), TN ($r = 0.45$, $p = 0.001$), TP ($r = 0.42$, $p = 0.001$) and AVN ($r = 0.23$, $p = 0.004$) had significant effects on the bacterial community structure (Figure 4C). For the fungal community structure, soil BD ($r = 0.17$, $p = 0.033$), TN ($r = 0.17$, $p = 0.016$), TK ($r = 0.21$, $p = 0.002$) and AVN ($r = 0.15$, $p = 0.032$) were the main factors.

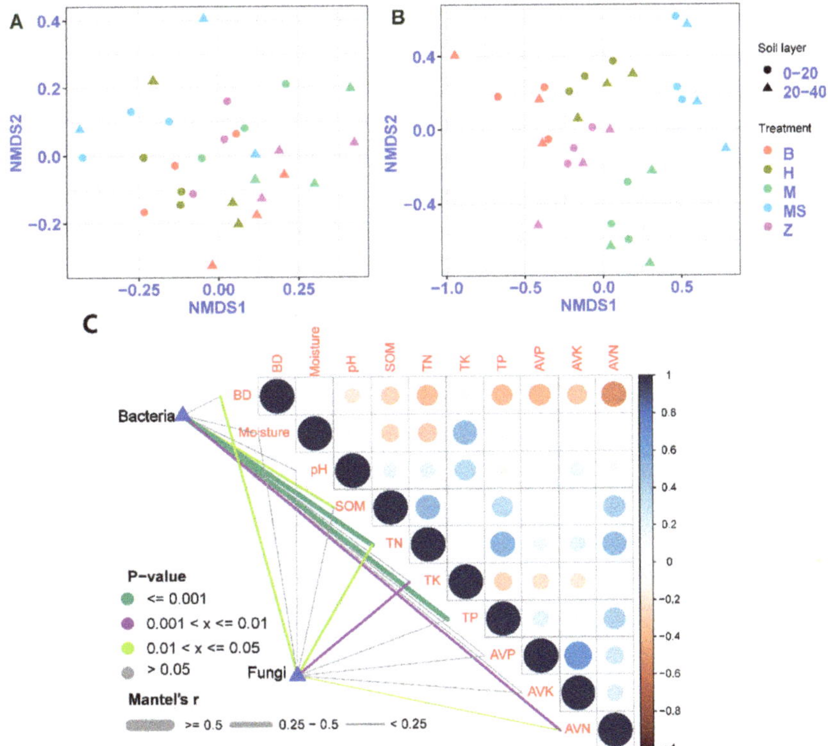

Figure 4. The community structure of soil bacteria (**A**) and fungi (**B**), and the associations between community structure and soil properties (**C**). B, *Brachiaria eruciformis*; Z, *Stylosanthes guianensis*; H, *Butterfly pea*; M and MS, no-till + herbicide. In the control (M and MS), soils were collected between the rows (M) and the drip line of mango trees (MS).

3.5. Variations in Soil Fungal Ecological Guilds under Different Sown Grass Strips

Soil functional fungal groups significantly altered under different cover cropping with different sown grass strips. Cover cropping enhanced the proportion of pathotroph fungi and decreased saprotroph fungi. Cover cropping increased the proportion of Arbuscular Mycorrhizal fungi and wood saprotroph fungi, while cover cropping had no significant effects on the plant pathogen or animal pathogen (Figure 5).

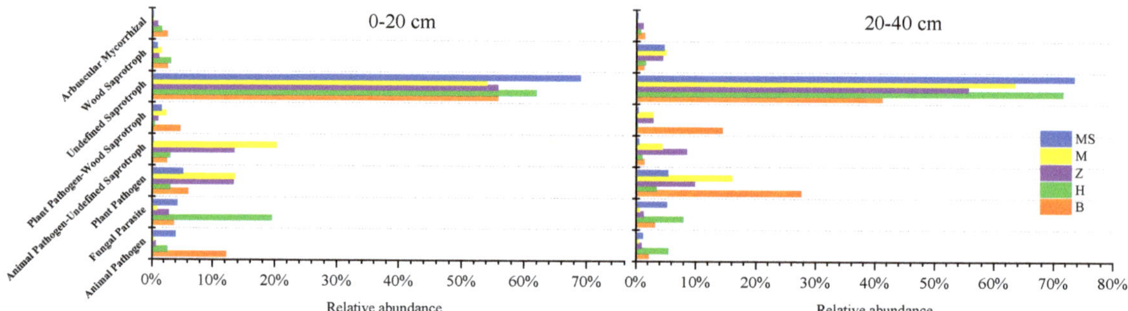

Figure 5. The functional groups of soil fungi under different cover cropping treatments. B, *Brachiaria eruciformis*; Z, *Stylosanthes guianensis*; H, *Butterfly pea*; M, no-till + herbicide. In the control, soils were collected between the rows (M) and the drip line of mango trees (MS).

3.6. The Associations between Soil Microbial Community and Diversity and Mango Fruit Yields and Quality

The correlation between fungal diversity (richness) and mango fruit yields was significant ($r = 0.75$, $p < 0.01$) (Figure 6). The soil fungal community structure (repressed by NMDS1) was significantly correlated with mango fruit yields ($r = -0.79$, $p < 0.01$). These also showed significant correlations between fungal diversity and mango fruit TSS ($r = 0.71$, $p = 0.01$) and organic acid ($r = -0.76$, $p < 0.01$). However, the soil bacterial diversity had no significant associations with mango fruit yields and TSS (data not shown).

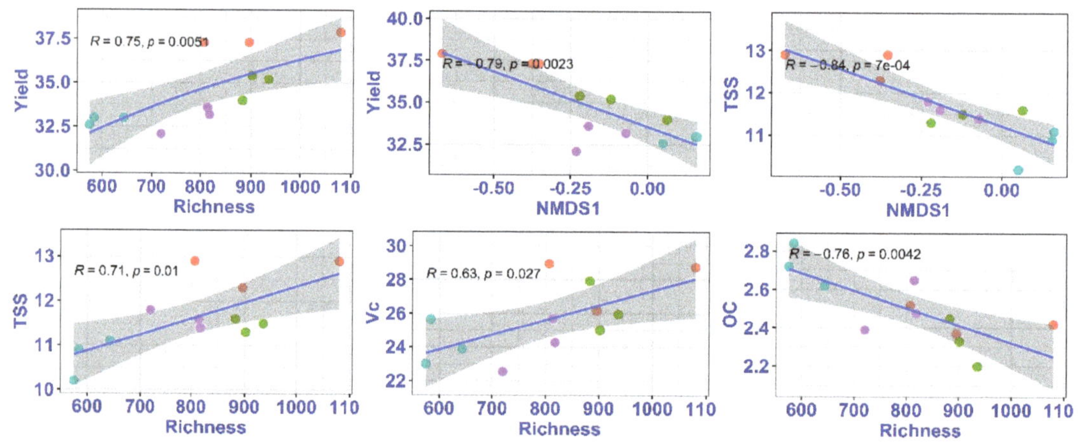

Figure 6. The associations between fungal community characteristics and mango fruit yield and quality under different cover cropping treatments. TSS, total soluble solids content; Vc, vitamin C; OC, organic acid. NMDS1 represented the community structure of soil fungi, and Richness indicated fungal diversity.

4. Discussion

Cover cropping has been widely used in agriculture systems to improve crop yield and quality as well as soil quality. In our study, the mango yield was significantly enhanced by 3–14% after the application of sown grass strips between rows. Similarly, winter cover cropping improved corn yields [39], and soybean yield significantly increased after the 3-year application with a multispecies mixture of legumes, grasses and Brassica spp [40]. These positive effects of cover cropping in crop yields suggested that cover cropping is an effective method to increase yields. In this study, we also found a significant increase in mango yields with B and H. However, Z had a slight increase in mango yields, suggesting that Z is not suitable to use in mango orchards. The positive effects of B and H on mango yields might be explained by the variations in nutrients and microbial community diversity in soils caused by cover cropping.

A meta-analysis found that global cropland soil carbon changes due to cover cropping, which increases SOC in near-surface soils by an average of 15.5% [2]. However, in our study, we found the sown grass did not enhance SOC in both soil layers. This difference might be explained by the duration of application of the sown grass. Sown grass strips caused the changes in available nutrients, such as AVN and AVP. The growth of sown grass could produce root exudates and litters, which were the main resources of soil nutrients, especially available nutrients. These variations in soil nutrients directly and indirectly influenced soil microbial diversity and community compositions, respectively.

In this study, we found that soil fungal diversity was sensitive to the application of sown grass strips. The introduction of sown grass between the rows enhanced the fungal diversity compared to the control (no sown grass). This might be explained by the higher decomposition ability of litters and roots than bacteria. Fungi are considered the primary decomposers of dead plant biomass in terrestrial ecosystems. The results from the litter-bag decomposition experiments in the field and the laboratory indicated the overwhelming advantage of fungi during the litter decomposition process [41–44]. Pascoal and Cássio (2004) showed that the contribution of fungi to litter decomposition greatly exceeded that of bacteria [45]. Despite the critic roles of fungi, the roles of bacteria could not be neglected [46], especially during the litter decomposition in which they mainly worked at the later decomposition stage. Therefore, the long-term introduction of sown grass strips might cause variations in soil bacterial diversity and community.

Different types of sown grasses also had different effects on soil fungal diversity. B and H soils had higher soil fungal diversity than Z. These variations might be explained by the quality of plant litter, roots and root exudates, which were the main factors impacting soil fungal diversity and community. For example, high quality litter decomposed faster than the low-quality litter [47,48]. The fast decomposition released a much higher amount of nutrients to the soil and resulted in the fast succession of soil fungi.

Nitrogen was considered as the main resource of soil microbes and strongly affected soil biodiversity and community structure [26,49]. Wang et al. (2018) found that tropical forest soil microbial community composition was shaped by N addition, with the increase in the proportion of arbuscular mycorrhizal fungi [50]. In the present study, we found that soil TN and AVN had significant effects on soil bacterial diversity, suggesting that soil N was the best predictor of the regulation of soil bacterial diversity in the mango orchards.

Soil microorganisms are considered as the main regulator of soil nutrient cycles and net primary production. In this study, we found that soil microbes enhanced mango fruit yields and quality via increasing fungal diversity and alterations in fungal community structure. Soil microbial diversity may improve crop yields through these mechanisms. First, soil biodiversity mediated nutrients available in the soil, which were the main resources of crop plant growth. Many previous studies have suggested that soil biodiversity enhances plant growth and drives crop yields [51,52]. Second, soil microorganisms help to maintain soil health and prevent the invasion of pathogens. Higher soil biodiversity provides higher ecosystem functions, such as net primary production and nutrient cycling. Therefore, soil microbial diversity, especially for fungal diversity, is vital in order to improve mango fruit yields.

5. Conclusions

This field experiment showed that cover cropping with sown grass had significant effects on the soil microbial community in a mango orchard. Fungal diversity was more sensitive than bacterial diversity in response to sown grass. The application of *Brachiaria eruciformis* (B) had the strongest effects on soil fungal diversity. The increase in fungal diversity suggested that sown grass had positive influences on soil biodiversity. Across soil properties, AVN and TN were the most important predictors affecting soil bacterial communities, while soil nutrients (TN and AVN) were the most important factors in mediating soil fungal communities. These results show that the sown grass strips in a mango orchard regulated the soil microbial community and diversity via soil organic matter and nitrogen, which might be of great significance in mango production and quality and the suitability of mango orchards. Mixtures of sown grasses might be more effective for soil biodiversity and soil function, and operations in mango orchards and future research should be focused on this aspect.

Author Contributions: Conceptualization, W.T. and Z.W.; methodology, Z.W.; data analysis, Q.Z. and Z.W.; investigation, Z.W.; writing—original draft preparation, Q.Z. and Z.W.; writing—review and editing, W.T.; visualization, W.T.; supervision, Z.W.; project administration, Z.W.; funding acquisition, W.T. All authors have read and agreed to the published version of the manuscript.

Funding: This research was supported by the National Natural Science Foundation of China (Nos. 41977023 and 32061123007).

Institutional Review Board Statement: Not applicable.

Informed Consent Statement: Not applicable.

Data Availability Statement: Data are contained within the article.

Conflicts of Interest: The authors declare no conflict of interest.

References

1. Olson, K.R.; Ebelhar, S.A.; Lang, J.M. Cover crop effects on crop yields and soil organic carbon content. *Soil Sci.* **2010**, *175*, 89–98. [CrossRef]
2. Jian, J.; Du, X.; Reiter, M.S.; Stewart, R.D. A meta-analysis of global cropland soil carbon changes due to cover cropping. *Soil Biol. Biochem.* **2020**, *143*, 107735. [CrossRef]
3. Van Eerd, L.L.; Congreves, K.A.; Hayes, A.; Verhallen, A.; Hooker, D.C. Long-term tillage and crop rotation effects on soil quality, organic carbon, and total nitrogen. *Can. J. Soil Sci.* **2014**, *94*, 303–315. [CrossRef]
4. Bandick, A.K.; Dick, R.P. Field management effects on soil enzyme activities. *Soil Biol. Biochem.* **1999**, *31*, 1471–1479. [CrossRef]
5. Lal, R. Soil carbon sequestration impacts on global climate change and food security. *Science* **2004**, *304*, 1623–1627. [CrossRef] [PubMed]
6. Zhou, J.; Xue, K.; Xie, J.; Deng, Y.; Wu, L.; Cheng, X.; Fei, S.; Deng, S.; He, Z.; Van Nostrand, J.D. Microbial mediation of carbon-cycle feedbacks to climate warming. *Nat. Clim. Chang.* **2012**, *2*, 106–110. [CrossRef]
7. Bloor, J.M.G.; Niboyet, A.; Leadley, P.W.; Barthes, L. CO_2 and inorganic N supply modify competition for N between co-occurring grass plants, tree seedlings and soil microorganisms. *Soil Biol. Biochem.* **2009**, *41*, 544–552. [CrossRef]
8. Elfstrand, S.; Hedlund, K.; Mårtensson, A. Soil enzyme activities, microbial community composition and function after 47 years of continuous green manuring. *Appl. Soil Ecol.* **2007**, *35*, 610–621. [CrossRef]
9. Berendsen, R.L.; Pieterse, C.M.J.; Bakker, P.A.H.M. The rhizosphere microbiome and plant health. *Trends Plant Sci.* **2012**, *17*, 478–486. [CrossRef]
10. Vasugi, C.; Dinesh, M.; Sekar, K.; Shivashankara, K.; Padmakar, B.; Ravishankar, K. Genetic diversity in unique indigenous mango accessions (Appemidi) of the Western Ghats for certain fruit characteristics. *Curr. Sci.* **2012**, *103*, 199–207.
11. Tharanathan, R.; Yashoda, H.; Prabha, T. Mango (*Mangifera indica* L.), "The king of fruits"—An overview. *Food Rev. Int.* **2006**, *22*, 95–123. [CrossRef]
12. Neguse, T.B.; Wanzala, F.K.; Ali, W.M.; Owino, W.O.; Mwangi, G.S. Mango (*Mangifera indica* L.) production practices and constraints in major production regions of Ethiopia. *Afr. J. Agric. Res.* **2019**, *14*, 185–196.
13. Massah, J.; Azadegan, B. Effect of chemical fertilizers on soil compaction and degradation. *Agric. Mech. Asia Africa Lat. Am.* **2016**, *47*, 44–50.

14. Fageria, N. Green manuring in crop production. *J. Plant Nutr.* **2007**, *30*, 691–719. [CrossRef]
15. Baggs, E.; Watson, C.; Rees, R. The fate of nitrogen from incorporated cover crop and green manure residues. *Nutr. Cycl. Agroecosyst* **2000**, *56*, 153–163. [CrossRef]
16. Garcia-Franco, N.; Albaladejo, J.; Almagro, M.; Martínez-Mena, M. Beneficial effects of reduced tillage and green manure on soil aggregation and stabilization of organic carbon in a Mediterranean agroecosystem. *Soil Tillage Res.* **2015**, *153*, 66–75. [CrossRef]
17. Mazzoncini, M.; Sapkota, T.B.; Barberi, P.; Antichi, D.; Risaliti, R. Long-term effect of tillage, nitrogen fertilization and cover crops on soil organic carbon and total nitrogen content. *Soil Tillage Res.* **2011**, *114*, 165–174. [CrossRef]
18. Sainju, U.M.; Singh, B.P.; Whitehead, W.F. Long-term effects of tillage, cover crops, and nitrogen fertilization on organic carbon and nitrogen concentrations in sandy loam soils in Georgia, USA. *Soil Tillage Res.* **2002**, *63*, 167–179. [CrossRef]
19. Duval, M.E.; Galantini, J.A.; Capurro, J.E.; Martinez, J.M. Winter cover crops in soybean monoculture: Effects on soil organic carbon and its fractions. *Soil Tillage Res.* **2016**, *161*, 95–105. [CrossRef]
20. Tolbert, V.R.; Todd, D.E.; Mann, L.K.; Jawdy, C.M.; Mays, D.A.; Malik, R.; Bandaranayake, W.; Houston, A.; Tyler, D.; Pettry, D.E. Changes in soil quality and below-ground carbon storage with conversion of traditional agricultural crop lands to bioenergy crop production. *Environ. Pollut.* **2002**, *116*, S97–S106. [CrossRef]
21. Wang, A.-T.; Wang, Q.; Li, J.; Yuan, G.-L.; Albanese, S.; Petrik, A. Geo-statistical and multivariate analyses of potentially toxic elements' distribution in the soil of Hainan Island (China): A comparison between the topsoil and subsoil at a regional scale. *J. Geochem. Explor.* **2019**, *197*, 48–59. [CrossRef]
22. Patil, A.S.; Maurer, D.; Feygenberg, O.; Alkan, N. Exploring cold quarantine to mango fruit against fruit fly using artificial ripening. *Sci. Rep.* **2019**, *9*, 1948. [CrossRef]
23. Lu, Y.; Si, B.; Li, H.; Biswas, A. Elucidating controls of the variability of deep soil bulk density. *Geoderma* **2019**, *348*, 146–157. [CrossRef]
24. Ren, H.; Xu, Z.; Huang, J.; Lü, X.; Zeng, D.-H.; Yuan, Z.; Han, X.; Fang, Y. Increased precipitation induces a positive plant-soil feedback in a semi-arid grassland. *Plant Soil* **2015**, *389*, 211–223. [CrossRef]
25. Thomas, R.; Sheard, R.; Moyer, J. Comparison of conventional and automated procedures for nitrogen, phosphorus, and potassium analysis of plant material using a single digestion 1. *Agron. J.* **1967**, *59*, 240–243. [CrossRef]
26. Zeng, Q.C.; An, S.S.; Liu, Y. Soil bacterial community response to vegetation succession after fencing in the grassland of China. *Sci. Total Environ.* **2017**, *609*, 2–10. [CrossRef] [PubMed]
27. Caporaso, J.G.; Lauber, C.L.; Walters, W.A.; Berglyons, D.; Lozupone, C.A.; Turnbaugh, P.J.; Fierer, N.; Knight, R. Global patterns of 16S rRNA diversity at a depth of millions of sequences per sample. *Proc. Natl. Acad. Sci. USA* **2011**, *108* (Suppl. 1), 4516–4522. [CrossRef]
28. Zeng, Q.; An, S. Identifying the Biogeographic Patterns of Rare and Abundant Bacterial Communities Using Different Primer Sets on the Loess Plateau. *Microorganisms* **2021**, *9*, 139. [CrossRef] [PubMed]
29. Fujita, S.-I.; Senda, Y.; Nakaguchi, S.; Hashimoto, T. Multiplex PCR Using Internal Transcribed Spacer 1 and 2 Regions for Rapid Detection and Identification of Yeast Strains. *J. Clin. Microbiol.* **2001**, *39*, 3617–3622. [CrossRef] [PubMed]
30. Zeng, Q.; Jia, P.; Wang, Y.; Wang, H.; Li, C.; An, S. The local environment regulates biogeographic patterns of soil fungal communities on the Loess Plateau. *Catena* **2019**, *183*, 104220. [CrossRef]
31. Ni, Y.; Yang, T.; Zhang, K.; Shen, C.; Chu, H. Fungal Communities Along a Small-Scale Elevational Gradient in an Alpine Tundra Are Determined by Soil Carbon Nitrogen Ratios. *Front. Microbiol.* **2018**, *9*, 1815. [CrossRef] [PubMed]
32. Delgado-Baquerizo, M.; Reich, P.B.; Trivedi, C.; Eldridge, D.J.; Abades, S.; Alfaro, F.D.; Bastida, F.; Berhe, A.A.; Cutler, N.A.; Gallardo, A. Multiple elements of soil biodiversity drive ecosystem functions across biomes. *Nat. Ecol. Evol.* **2020**, *4*, 210–220. [CrossRef]
33. Edgar, R.C.; Haas, B.J.; Clemente, J.C.; Quince, C.; Knight, R. UCHIME improves sensitivity and speed of chimera detection. *Bioinformatics* **2011**, *27*, 2194. [CrossRef] [PubMed]
34. Nguyen, N.H.; Song, Z.; Bates, S.T.; Branco, S.; Tedersoo, L.; Menke, J.; Schilling, J.S.; Kennedy, P.G. FUNGuild: An open annotation tool for parsing fungal community datasets by ecological guild. *Fungal Ecol.* **2016**, *20*, 241–248. [CrossRef]
35. Větrovský, T.; Kohout, P.; Kopecký, M.; Machac, A.; Man, M.; Bahnmann, B.D.; Brabcová, V.; Choi, J.; Meszárošová, L.; Human, Z.R.; et al. A meta-analysis of global fungal distribution reveals climate-driven patterns. *Nat. Commun.* **2019**, *10*, 5142. [CrossRef]
36. Delgado-Baquerizo, M.; Guerra, C.A.; Cano-Díaz, C.; Egidi, E.; Wang, J.-T.; Eisenhauer, N.; Singh, B.K.; Maestre, F.T. The proportion of soil-borne pathogens increases with warming at the global scale. *Nat. Clim. Chang.* **2020**, *10*, 550–554. [CrossRef]
37. Oksanen, J.; Blanchet, F.G.; Kindt, R.; Legendre, P.; Minchin, P.R.; O'hara, R.; Simpson, G.L.; Solymos, P.; Stevens, M.H.H.; Wagner, H. Package 'vegan'. *Community Ecol. Package Version* **2013**, *2*, 1–295.
38. Goslee, S.C.; Urban, D.L. The ecodist package for dissimilarity-based analysis of ecological data. *J. Stat. Softw.* **2007**, *22*, 1–19. [CrossRef]
39. Kuo, S.; Sainju, U.; Jellum, E. Winter cover cropping influence on nitrogen mineralization, presidedress soil nitrate test, and corn yields. *Biol. Fertil. Soils* **1996**, *22*, 310–317. [CrossRef]
40. Chu, M.; Jagadamma, S.; Walker, F.R.; Eash, N.S.; Buschermohle, M.J.; Duncan, L.A. Effect of multispecies cover crop mixture on soil properties and crop yield. *Agric. Environ. Lett.* **2017**, *2*, 170030. [CrossRef]

41. Stursova, M.; Zifcakova, L.; Leigh, M.B.; Burgess, R.; Baldrian, P. Cellulose utilization in forest litter and soil: Identification of bacterial and fungal decomposers. *FEMS Microbiol. Ecol.* **2012**, *80*, 735–746. [CrossRef] [PubMed]

42. Aerts, R. Climate, leaf litter chemistry and leaf litter decomposition in terrestrial ecosystems: A triangular relationship. *Oikos* **1997**, *79*, 439. [CrossRef]

43. Voříšková, J.; Baldrian, P. Fungal community on decomposing leaf litter undergoes rapid successional changes. *ISME J.* **2012**, *7*, 477–486. [CrossRef] [PubMed]

44. Ferreira, V.; Chauvet, E. Synergistic effects of water temperature and dissolved nutrients on litter decomposition and associated fungi. *Glob. Chang. Biol.* **2011**, *17*, 551–564. [CrossRef]

45. Pascoal, C.; Cássio, F. Contribution of fungi and bacteria to leaf litter decomposition in a polluted river. *Appl. Environ. Microbiol.* **2004**, *70*, 5266–5273. [CrossRef]

46. Zeng, Q.; Liu, Y.; Zhang, H.; An, S. Fast bacterial succession associated with the decomposition of Quercus wutaishanica litter on the Loess Plateau. *Biogeochemistry* **2019**, *144*, 119–131. [CrossRef]

47. Zhang, P.; Tian, X.; He, X.; Song, F.; Ren, L.; Jiang, P. Effect of litter quality on its decomposition in broadleaf and coniferous forest. *Eur. J. Soil Biol.* **2008**, *44*, 392–399. [CrossRef]

48. Cuchietti, A.; Marcotti, E.; Gurvich, D.E.; Cingolani, A.M.; Harguindeguy, N.P. Leaf litter mixtures and neighbour effects: Low-nitrogen and high-lignin species increase decomposition rate of high-nitrogen and low-lignin neighbours. *Appl. Soil Ecol.* **2014**, *82*, 44–51. [CrossRef]

49. Deng, Q.; Cheng, X.; Hui, D.; Zhang, Q.; Li, M.; Zhang, Q. Soil microbial community and its interaction with soil carbon and nitrogen dynamics following afforestation in central China. *Sci. Total. Environ.* **2016**, *541*, 230–237. [CrossRef]

50. Wang, C.; Lu, X.; Mori, T.; Mao, Q.; Zhou, K.; Zhou, G.; Nie, Y.; Mo, J. Responses of soil microbial community to continuous experimental nitrogen additions for 13 years in a nitrogen-rich tropical forest. *Soil Biol. Biochem.* **2018**, *121*, 103–112. [CrossRef]

51. Saleem, M.; Hu, J.; Jousset, A. More than the sum of its parts: Microbiome biodiversity as a driver of plant growth and soil health. *Annu. Rev. Ecol. Evol. Syst.* **2019**, *50*, 145–168. [CrossRef]

52. Rillig, M.C.; Lehmann, A.; Lehmann, J.; Camenzind, T.; Rauh, C. Soil Biodiversity Effects from Field to Fork. *Trends Plant Sci.* **2018**, *23*, 17–24. [CrossRef] [PubMed]

MDPI
St. Alban-Anlage 66
4052 Basel
Switzerland
www.mdpi.com

Agriculture Editorial Office
E-mail: agriculture@mdpi.com
www.mdpi.com/journal/agriculture